155 Topics in Current Chemistry

Small Ring Compounds in Organic Synthesis IV

Editor: A. de Meijere

With contributions by
T.-L. Ho, H. Hopf, R. R. Kostikov, I. Kuwajima,
A. P. Molchanov, E. Nakamura

With 16 Tables

Springer-Verlag Berlin Heidelberg GmbH

This series presents critical reviews of the present position and future trends in modern chemical research. It is addressed to all research and industrial chemists who wish to keep abreast of advances in their subject.

As a rule, contributions are specially commissioned. The editors and publishers will, however, always pleased to be receive suggestions and supplementary information. Papers are accepted for "Topics in Current Chemistry" in English.

ISBN 978-3-662-15015-3   ISBN 978-3-540-46984-1 (eBook)
DOI 10.1007/978-3-540-46984-1

© Springer-Verlag Berlin Heidelberg 1990
Originally published by Springer-Verlag Berlin Heidelberg New York in 1990
Softcover reprint of the hardcover 1st edition 1990

2151/3020-543210 — Printed on acid-free paper

Guest Editor

Prof. Dr. *Armin De Meijere*
Universität Hamburg
Institut f. Org. Chemie u. Biochemie
Martin-Luther-King-Platz 6
D-2000 Hamburg 13

Editorial Board

Table of Contents

Metal Homoenolates from Siloxycyclopropanes

Isao Kuwajima and Eiichi Nakamura

Department of Chemistry, Tokyo Institute of Technology, Meguro, Tokyo, Japan 152

Table of Contents

The potential of metal homoenolates as multifunctional reagents in organic synthesis has become recognized in the last decade. Ring opening of siloxycyclopropanes by Lewis acidic metals currently provides the best route to the metal homoenolates, which are either isolable as a stable complex or exist only as a transient organometallic species. This review summarizes, for the first time, the research activities of metal homoenolate chemistry. Other synthetically useful routes to metal homoenolates are also described.

1 Introduction

Although the chemistry of metal enolate/α-metallo ketone tautomeric pairs Eq. (1) has attracted the interest of chemists for a long period, it still represents an important subject of organic chemistry. Recent vigorous activities in achieving stereocontrol in the aldol reaction, for instance, resulted in further refinement of enolate chemistry. Unlike enolates, the homoenolate tautomeric pair Eq. (2) [1] in which the carbonyl group and the β-anionic center are homoconjugated, has eluded detailed investigation, in particular, that of synthetic applications [2].

$$\text{(1)}$$

$$\text{(2)}$$

Like that of an enolate, the value of a homoenolate stems from its amphoteric nature (Cf. Eqs. 1, 2). In addition, the homoenolate serves as an archetypal synthon in the concept of Umpolung [3], acting as a nucleophilic, inverse-polarity Michael acceptor. Despite such conceptual importance, practical problems set great restrictions on the actual application of homoenolate chemistry in organic synthesis. First, unlike the enolate anion, the homoenolate anion cannot be stoichiometrically generated by deprotonation of a carbonyl compound, since the pK_a value of the β-hydrogen is only very slightly lowered [2]. An additional problem is created by an intrinsic property of reactive homoenolates; namely sodium or lithium homoenolates spontaneously cyclize to the cyclopropanolate tautomer [2]. This alkoxide tautomer has never reacted as a carbon-nucleophile toward standard electrophiles (except a proton). Therefore, in the subsequent paragraphs, the word homoenolate refers specifically to the carbanionic form of the homoenolate tautomeric pair Eq. (2). Transition metal homoenolates have been postulated in the reactions of acyl transition metal complexes with olefins, which is a common process in reactions in a carbon monoxide atmosphere Eq. (3) [4]. However, without suitable precautions to stabilize the complex [5], these transition metal homoenolates undergo further reactions.

$$\text{(3)}$$

Nonetheless, "homoenolates" have been much talked about as a species useful for carbon-carbon chain extension, and a number of "homoenolate equivalents" have been prepared and used for organic synthesis. Reviews on these species have been published [6].

The last few years have seen the development of an entirely new methodology in which the metal homoenolate is prepared by ring opening of siloxycyclopropanes with metal halides Eq. (4). In these reactions, a very subtle balance of kinetic and thermodynamic factors allows certain metal halides to form reactive metal homoenolates which do not close back to the three-membered ring. Although the reaction in Eq. (4) tolerates a broad spectrum of structural variation with respect to the substituents, the nature of the substituent on the ring, particularly that at C-1, has a great influence on the chemistry following the ring cleavage. 1-Alkoxy substituents (R^1) exert particularly great effects, allowing the preparation of a variety of ester homoenolates. Cyclopropanes with R^1 = alkyl or aryl thus far remain less viable as precursors to the corresponding homoenolates.

$$\text{structure} + MX_n \longrightarrow \text{structure} + XSi\Sigma \qquad (4)$$

The chemistry of cyclopropanol [7] has long been studied in the context of electrophilic reactions, and these investigations have resulted in the preparation of some 3-mercurio ketones. As such mercury compounds are quite unreactive, they have failed to attract great interest in homoenolate chemistry. Only recent studies to exploit siloxycyclopropanes as precursors to homoenolates have led to the use of 3-mercurio ketones for the transition metal-catalyzed formation of new carbon-carbon bonds [8] (vide infra).

Scheme 1

In 1977, an article from the authors' laboratories [9] reported an TiCl$_4$-mediated coupling reaction of 1-alkoxy-1-siloxy-cyclopropane with aldehydes (Scheme 1), in which the intermediate formation of a titanium homoenolate (path b) was postulated instead of a then-more-likely Friedel-Crafts-like mechanism (path a). This finding some years later led to the isolation of the first stable metal homoenolate [10] that exhibits considerable nucleophilic reactivity toward (external) electrophiles. Although the metal-carbon bond in this titanium complex is essentially covalent, such titanium species underwent ready nucleophilic addition onto carbonyl compounds to give 4-hydroxy esters in good yield. Since then a number of characterizable metal homoenolates have been prepared from siloxycyclopropanes [11]. The repertoire of metal homoenolate reactions now covers most of the standard reaction types ranging from simple elimination to substitution reactions. Homoenolates are now a reality in organic synthesis rather than a presumed intermediate of exotic reactivity. The purpose of this review is to summarize the current stage of development of the siloxycyclopropane route to metal homoenolates and convey information to chemists engaged in organic synthesis as well as organometallic chemistry. A few other recent routes to metal homoenolates are also reviewed briefly.

2 Preparation of Siloxycyclopropanes

In 1972 and 1973, four laboratories [12–15] reported the Simmons-Smith cyclopropanation of enol silyl ethers as a very flexible route to siloxycyclopropanes, Eq. (5). Reagents prepared either from a zinc/copper couple [13—15] or a zinc/ silver couple [12] have been successfully used for the cyclopropanation of enol silyl ethers from ketones (Table 1). Regioselectivity is noted with polyolefinic substrates, the electron-rich enol silyl ether moiety being attacked preferentially. The reaction is stereospecific with retention of the double bond configuration in the starting material. Prolonged heating of the reaction mixture results in a Zn(II)-catalyzed rearrangement of siloxycyclopropanes to allylic ethers. Termination of the cyclopropanation by addition of pyridine, aqueous NH_4Cl [15], or gaseous ammonia [16] prevents side reactions which may otherwise occur during workup owing to the Zn(II) residues. Table I shows some representative types of compounds prepared by this carbenoid route.

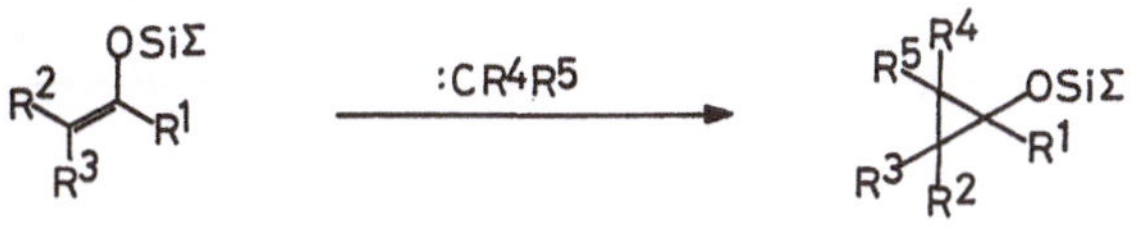

$$(5)$$

Table 1. Siloxycyclopropanes prepared by the Simmons-Smith reaction

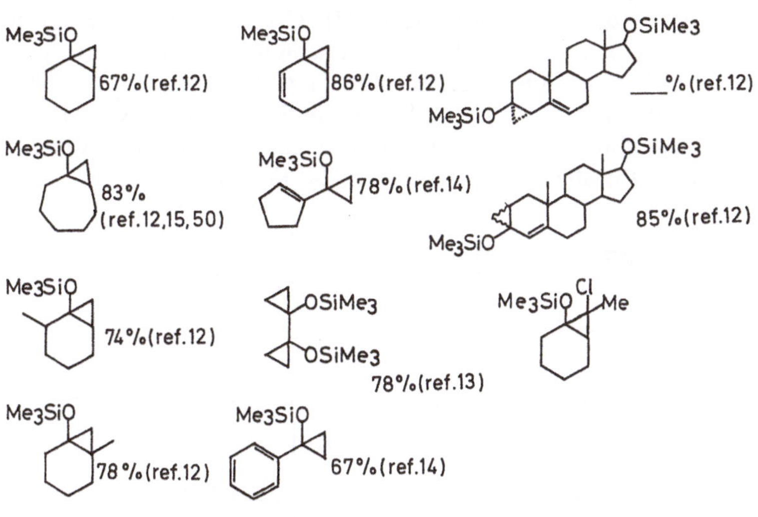

The carbenoid from Et_2Zn/CH_2I_2 [17], particularly when generated in the presence of oxygen [18], is more reactive than the conventional Simmons-Smith reagents. The milder conditions required are suitable for the preparation of 1-[16, 19] or 2-alkoxy-1-siloxycyclopropanes [20], which are generally more sensitive than the parent alkyl substituted siloxycyclopropanes (Table 2). Cyclopropanation of silyl ketene acetals is not completely stereospecific, since isomerization of the double bond in the starting material competes with the cyclopropanation [19].

Table 2. Alkoxy siloxy cyclopropanes prepared by Et_2Zn/CH_2I_2.

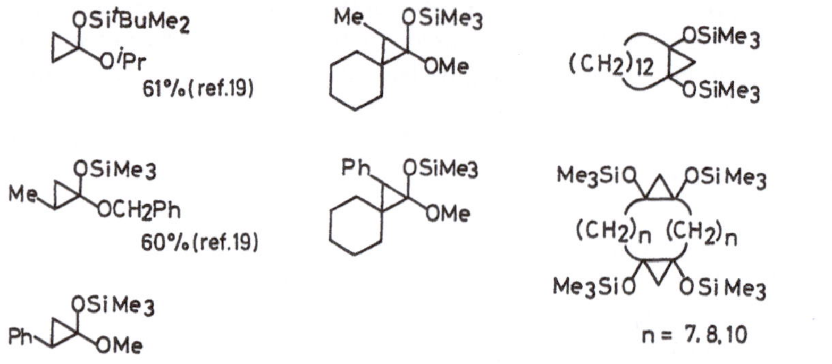

Acyloin-type reactions of esters provide the simplest route to 1-siloxy-1-alkoxycyclopropane [21, 22] Eq. (6). The reaction of commercial 3-halopropionate with sodium (or lithium) in refluxing ether in the presence of Me_3SiCl can easily be carried out on a one mole scale [21]. Cyclization of optically pure methyl 3-bromo-2-methylpropionate [23], available in both R and S form, gives a cyclopropane, which is enantiomerically pure at C-2, yet is a 1:1 diastereomeric mixture with respect to its relative configuration at C-1 Eq. (7). Reductive silylation of allyl 3-iodopropionate with zinc/copper couple provides a milder alternative to the alkali metal reduction [24] Eq. (8).

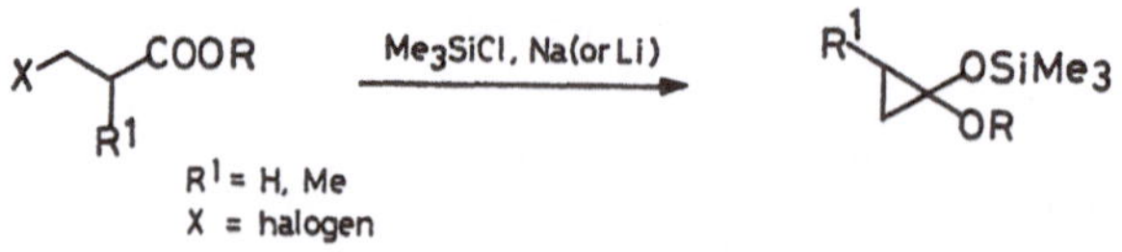

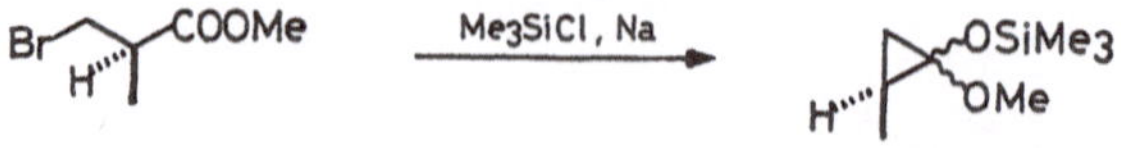

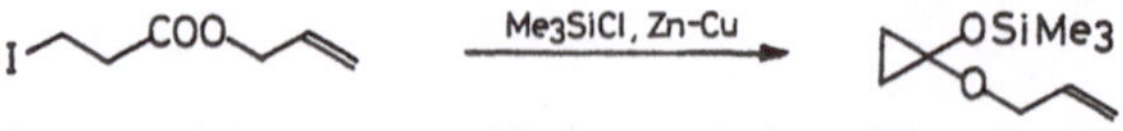

3 Preparation and Reactions of Isolable Metal Homoenolates

The recent surge of interests in metal homoenolate chemistry has been stimulated by the recognition that the siloxycyclopropane route can afford novel reactive homoenolate species that are stable enough for isolation, purification, and characterization. The stability of such homoenolates crucially depends on the subtle balance of nucleophilic and electrophilic reactivity of the two reactive sites in the molecule. Naturally, homoenolates with metal-carbon bonds that are too stable do not serve as nucleophiles in organic synthesis.

3.1 Preparation

3.1.1 Group 4 Metal Homoenolates

An exothermic reaction between cyclopropane 1 and $TiCl_4$ in methylene chloride produces a wine-red solution of a mixture of titanium homoenolate 2 and chlorotrimethylsilane [10, 19]. When the reaction is performed in hexane, the titanium species precipitates in the form of fine violet needles (approx 90% isolated yield, Eq. (9)).

1a : R = Et
1b   R = iPr

70~89%

This complex, while sensitive to oxygen and moisture, is stable for extended periods either neat or in solution. Its carbonyl stretching band (approx. 1610 cm$^-$) in the IR spectrum in dilute solution clearly indicates a chelate structure, which endows the carbon-titanium bond of this complex with very high stability, as compared with that of simple alkyltrichlorotitaniums [25].

Scheme 2

The chemical reactivities of such titanium homoenolates are similar to those of ordinary titanium alkyls (Scheme 2). Oxidation of the metal-carbon bond with bromine or oxygen occurs readily. Transmetalations with other metal halides such as $SnCl_4$, $SbCl_5$, $TeCl_4$, and $NbCl_5$ proceed cleanly. Reaction with benzaldehyde gives a 4-chloroester as the result of carbon-carbon bond formation followed by chlorination [9]. Acetone forms an addition complex. No reaction takes place with acid chloride and *tert*-alkyl chlorides.

The reaction of the cyclopropane 1 with $ZrCl_4$ gave a complex mixture [11].

3.1.2 Group 12 Metal Homoenolates

Cyclopropanols react with mercuric salts to give a variety of 3-mercurio ketones, and the chemistry of this ring cleavage has been thouroughly studied from the mechanistic viewpoint [7].

Similarly, siloxycyclopropanes (e.g., 3) react smoothly with mercuric acetate in methanol at 20 °C to give 3-mercurio ketones (e.g., 4) [8]. The attack of the metal occurs in such a way that the less alkyl-substituted bond is cleaved. Thus, starting from the enol silyl ether, the overall sequence represents α-mercurio-methylation of the parent ketone. The reaction is likely to proceed via an ionic intermediate Eq. (10).

Isomerization of siloxycyclopropane 3 to the allylic silyl ether in the presence of ZnI_2 has close mechanistic similarity to the homoenolate formation, Eq. (11) [26]. The initial zwitterionic intermediate after ring cleavage undergoes a hydride shift rather than elimination of the silyl group.

1-Alkoxy-1-siloxycyclopropanes 1 are also cleaved readily by $HgCl_2$, $Hg(OAc)_2$, or $Hg(OCOCF_3)_2$ to give monoalkylated mercuriaer 5 in high yield Eq. (12) [11]. Further alkylation of 5 was not observed. The methyl-substituted cyclopropane 6 produces 3-mercurio-2-methylpropionate as a single product, while the phenyl-substituted 7 gives a mixture of 2- and 3-phenylesters [27]. $CdCl_2$ reacts slowly in $CDCl_3$ under ultrasound irradiation at room temperature to give the cadmium homoenolate [11].

ZnCl$_2$ cleanly reacts with cyclopropane **1** in ether within a few hours at room temperature to give the zinc homoenolate etherate **8** in high yield Eq. (13) [11, 28]. Removal of the solvent gives the ether-free complex **9**, which can also be generated albeit, in lower yield, in a halomethane solvent. The zinc homoenolate, though sensitive to oxygen and water, is thermally very stable. The ring cleavage reaction is in fact an equilibrium, which in ether favors the homoenolate formation (K = approx 30) [29].

Methylcyclopropane **6** reacts selectively (>98%) with ZnCl$_2$ at the less hindered site to give the homoenolate of 2-methylpropionate, whereas phenyl substituted analogue **7** is cleaved selectively (>98%) at the more substituted site [27]. The latter homoenolate however suffers in situ protonation under the reaction conditions and could not be isolated.

The optically active cyclopropane **10** gave the chiral homoenolate of isobutyrate Eq. (14), an ethereal solution of which is both chemically and configurationally stable for a week at room temperature, and consequently can serve as a useful chiral building block [23].

3.1.3 Group 13 Metal Homoenolates

The common Lewis acids in this group, AlCl$_3$ and BX$_3$, do not form metal homoenolates in their reaction with siloxycyclopropanes, only GaCl$_3$ reacts with 1-alkoxy-1-siloxycyclopropanes **1** to give propionate homoenolates [11].

3.1.4 Group 14 Metal Homoenolates

Upon addition of cyclopropane **1** to one equivalent of SnCl$_4$ in methylene chloride at 0 °C, an exothermic reaction occurs to give 3-stannyl propionate **11** in approx. 70% yield [9, 11]. Addition of a second equiv of the cyclopropane slowly converts **11** to the dialkylated tin compounds, **12** Eq. (15). No further alkylation was observed. Chelate structures of such homoenolates implied by their spectral properties are supported by single crystal X-ray analysis [30].

The cyclopropane is inert to tributyltin chloride, but reacts readily with more highly Lewis acidic tributyltin trifluoromethanesulfonate (triflate) [11]. The 3-stannyl ester **13** showed no sign of internal chelation Eq. (16).

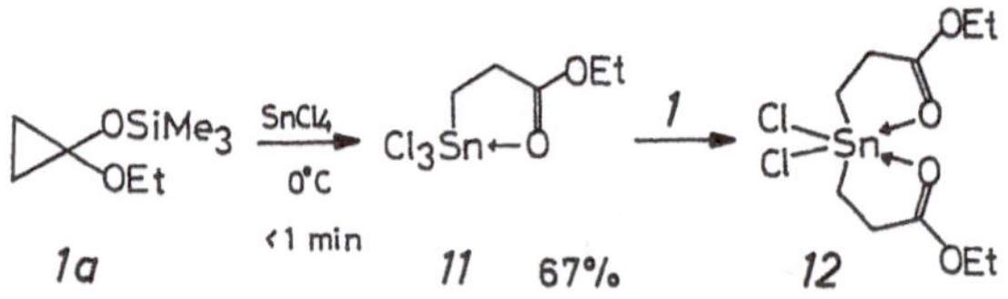

$$(15)$$

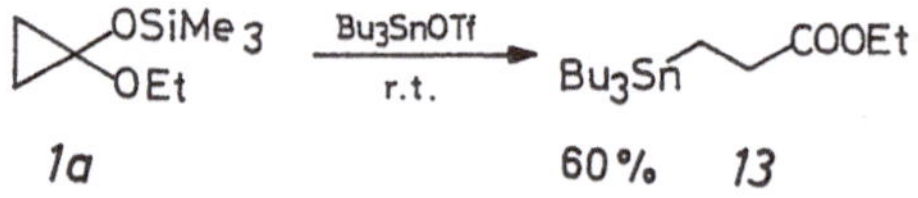

$$(16)$$

The cyclopropane **1** was inert to GeCl$_4$, SiCl$_4$, Me$_3$SiCl, trimethylsilyl triflate, and PbCl$_2$ [11].

Siloxycylopropanes corresponding to ketone homoenolates (e.g. **3**) also react smoothly with SnCl$_4$ at 15 °C to give 3-stannyl ketones (e.g. **14**) Eq. (17) [31]. In the same manner, the 3-stannyl aldehyde **15** has been prepared in good yield. Eq. (18).

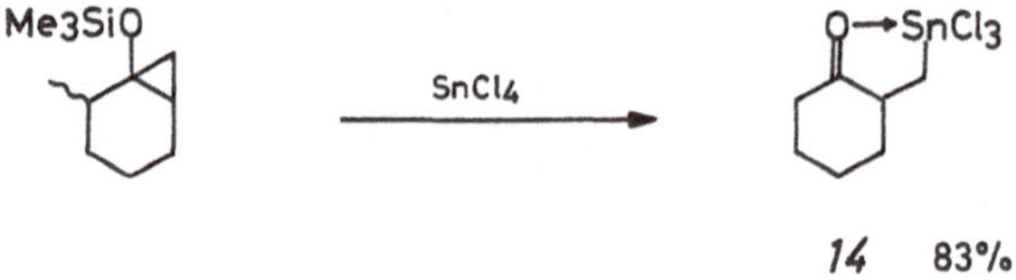

$$(17)$$

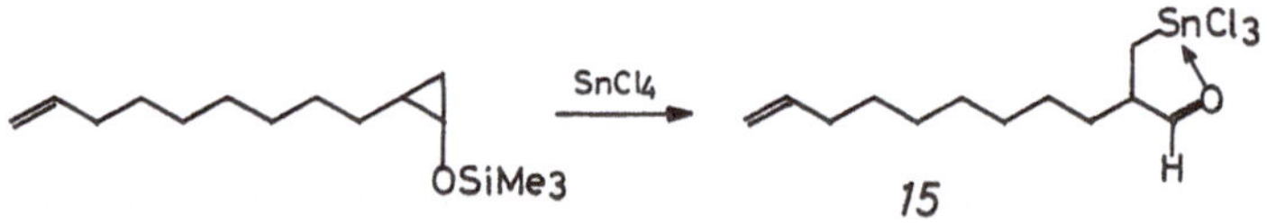

$$(18)$$

3.1.5 Group 15 Metal Homoenolates

Treatment of cyclopropane **1** with one equivalent of dry BiCl$_3$ in methylene chloride results in an exothermic reaction producing monoakylbismuth derivative **16** in 80% yield Eq. (19) [11]. Addition of another equivalent of the cyclopropane then affords the dialkylated bismuth species **17**, which in turn reacts with BiCl$_3$ to give the monoalkyl species. IR absorptions due to the carbonyl groups indicate the chelate structures shown. The two propionate moieties in the dialkylated compound **17** give rise to two distinctive carbonyl bands in the IR spectrum, but are indistinguishable by ^{1}H NMR, indicating rapid internal ligand exchange at 35 °C.

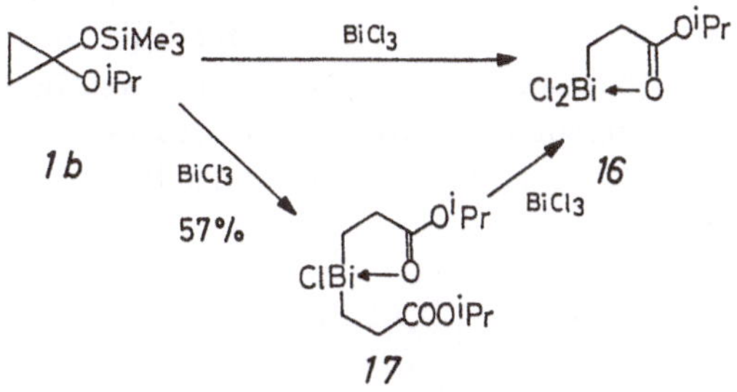

$$(19)$$

1b   BiCl₃
57%

16

17

Strongly Lewis acidic SbCl$_5$ rapidly reacts with one equivalent of the cyclo-propane **16** in chloroform to give monoalkylated antimony derivate **18** in 87% yield Eq. (20) [11]. The internal chelation is especially strong, as indicated by the carbonyl band in the IR spectrum at $1600\ cm^{-1}$. In contrast to the common monoalkyltetrachloroantimony, the chelated homoenolate **18** is stable at room temperature for many hours.

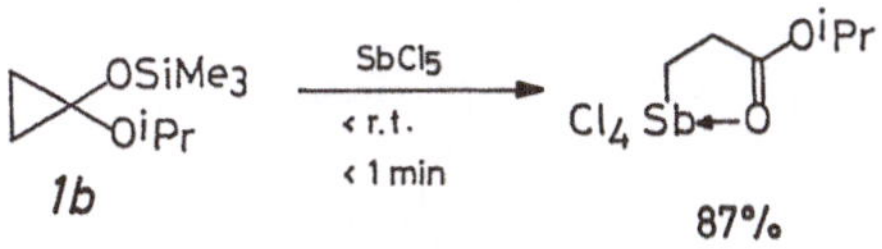

$$(20)$$

1b

87%

3.1.6 Group 16 Metal Homoenolates

Among the group 16 elements, mostly non-metallic ones like TeCl$_4$ were examined [11]. The reaction with TeCl$_4$ proceeds in a stepwise manner giving the mono-alkyltellurium and the dialkyl compound in accordance with the stoichiometry of the reagents Eq. (21). As in the case of tin and bismuth, the second alkylation is much slower than the first and the dialkyl species could not be alkylated any further.

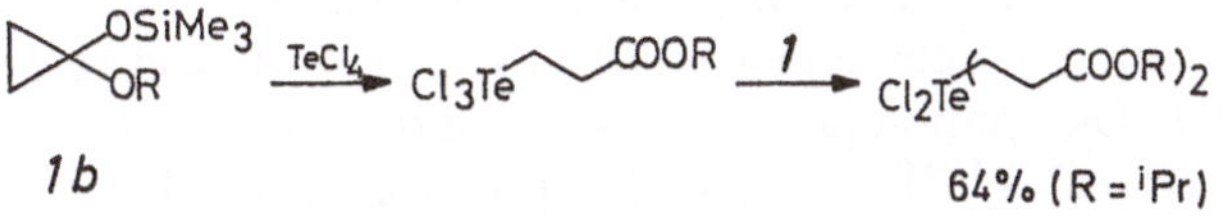

$$(21)$$

1b

64% (R = ^{i}Pr)

3.1.7 Metal Homoenolates of Other Groups [11]

The cyclopropane **1** reacts with none of the group 1 and 2 metal chlorides. Among early transition metal chlorides, NbCl$_4$ reacted with **1** in moderate yield to give the same homoenolate obtained by the reaction of equimolar amounts of titanium homoenolate **2** and NbCl$_4$ (Scheme 2). TaCl$_5$, CrCl$_3$, MoCl$_5$, and WCl$_5$ did not give any characterizable products.

3.2 Carbon-carbon Bond Forming Reactions of Metal Homoenolates

Various types of carbon-carbon bond forming reactions of metal homoenolates have been reported, some of which are highly synthetically useful. Scheme 3 illustrates reaction types of zinc homoenolates (**8** or **9**). In this section the reactions of stable homoenolates are presented according to the reaction types. Examples, in which siloxycyclopropanes generate transient, unstable homoenolates, are described in Sect. 5.

Scheme 3

3.2.1 Cyclopropane Formation

Internal nucleophilic cyclization leading to alkoxycyclopropanes is the most typical reaction of reactive metal homoenolates [1, 2]. The nature of stable homoenolates, however, is such that the anionic carbon C-3 forms a covalent bond with the metal atom and does not show high nucleophilic reactivity.

Among isolable metal homoenolates only zinc homoenolates cyclize to cyclopropanes under suitable conditions. Whereas acylation of zinc alkyls makes a straightforward ketone synthesis [32], that of a zinc homoenolate is more complex. Treatment of a purified zinc homoenolate in CDCl$_3$ with acid chloride at room temperature gives O-acylation product, instead of the expected 4-keto ester, as the single product (Eq. (22) [33]). The reaction probably proceeds by initial electrophilic attack of acyl cation on the carbonyl oxygen. A C-acylation leading to a 4-keto ester can, however, be accomplished in a polar solvent Eq. (44).

(22)

Treatment of zinc homoenolates with Me_3SiCl in a polar solvent also results in cyclopropane formation Eq. (23). This provides a very mild route to the siloxycyclopropanes [24].

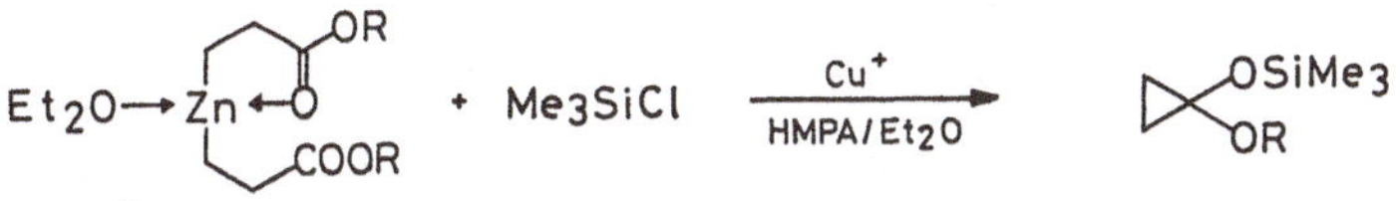

$$\tag{23}$$

3.2.2 Elimination

Next to the cyclopropane formation, elimination represents the simplest type of a carbon-carbon bond formation in the homoenolates. Transition metal homoenolates readily eliminate a metal hydride unit to give α,β-unsaturated carbonyl compounds. Treatment of a mercurio ketone with palladium (II) chloride results in the formation of the enone presumably via a 3-palladio ketone (Eq. (24), Table 3) [8]. The reaction can be carried out with catalytic amounts of palladium (II) by using $CuCl_2$ as an oxidant. Isomerization of the initial exomethylene derivative to the more stable *endo*-olefin can efficiently be retarded by addition of triethylamine to the reaction mixture.

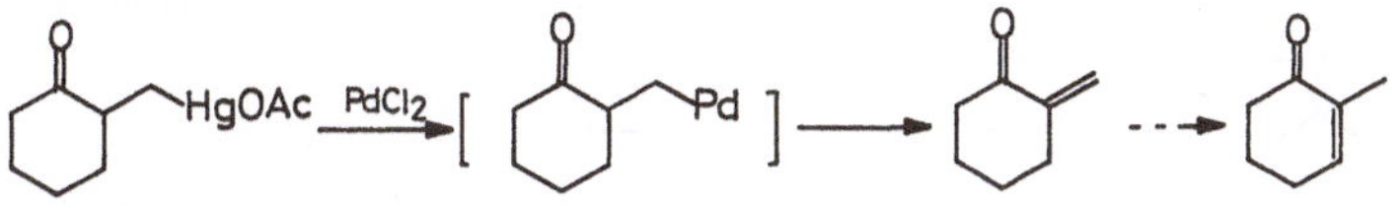

$$\tag{24}$$

Heating of the 3-trichlorostannyl ketone in DMSO also results in formation of 2-methylencyclohexanone (Eq. (25), Table 4) [31], which is formally the reverse reaction of a hydrostannylation of the enone with $HSnCl_3$. This alternative route

Table 3. Synthesis α-methylene ketones via mercurio ketones (Ref. [8])

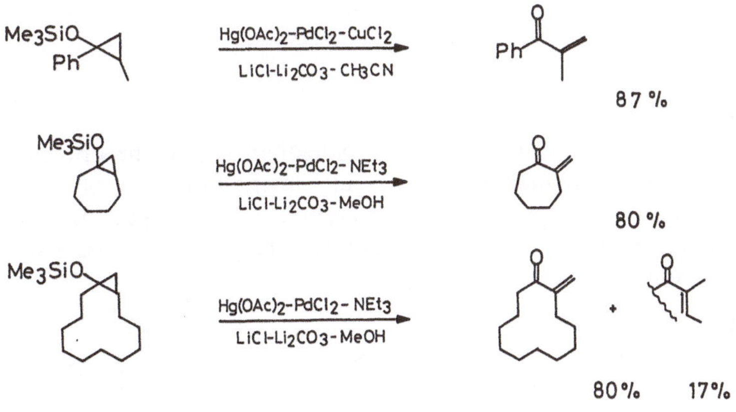

to α-methylene ketones has an obvious advantage over the above mentioned mercurial-palladium approach, but the overall reaction conditions of the latter are essentially neutral. The utility of the stannyl ketone route is illustrated by the preparation of an unstable α-methylene aldehyde, Eq. (26).

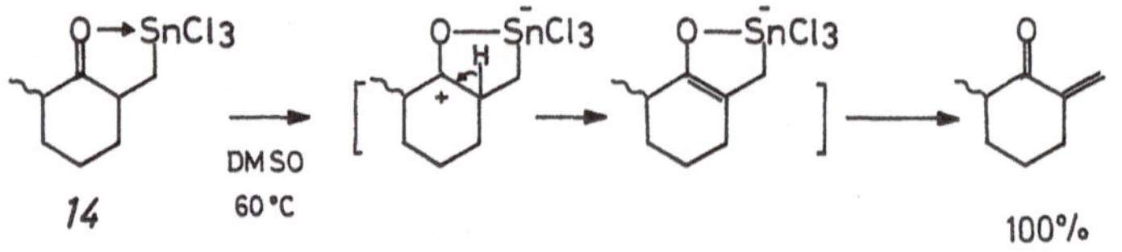

$$(25)$$

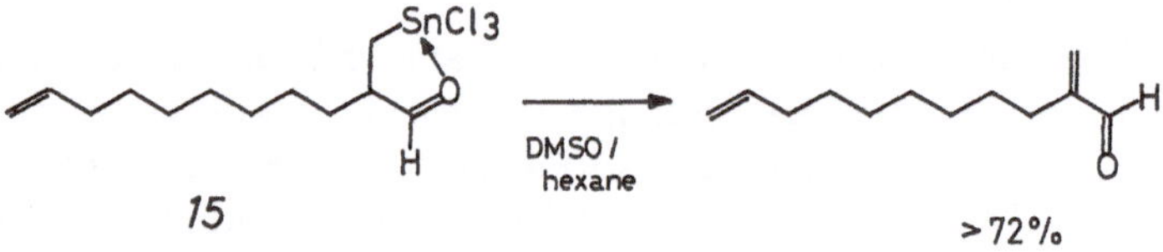

$$(26)$$

Table 4. Synthesis of α-methylene ketones from stannyl ketones (Rev. [3])

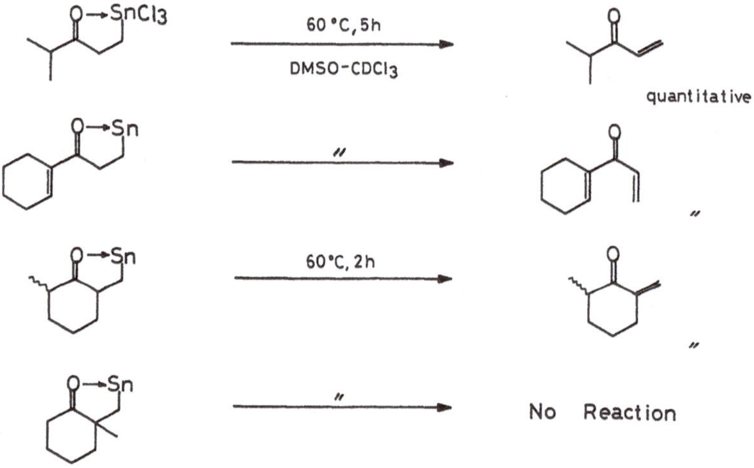

3.2.3 Carbonylation

The treatment of the mercurio ketone with palladium (II) in the presence of carbon monoxide and methanol, Eq. (27) results in the formation of a γ-keto ester with incorporation of one molecule of carbon monoxide [8]. The overall conversion of a siloxycyclopropane to the keto ester may be performed without isolation of the mercurio ketone.

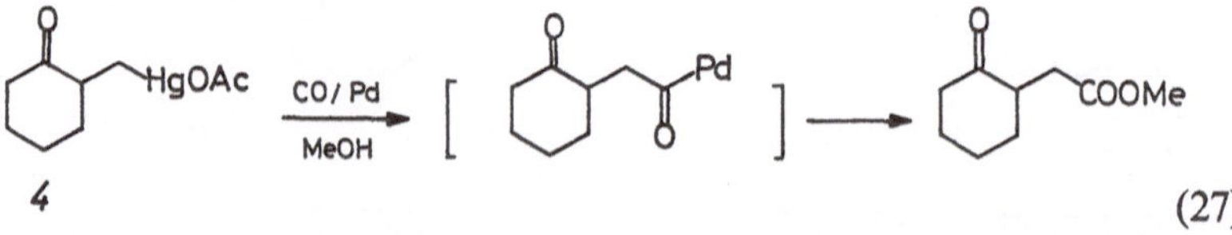

$$(27)$$

14

Treatment of the mercurio ketone with $Ni(CO)_4$ results in a symmetrical coupling with incorporation of one molecule of carbon monoxide to give a triketone presumably via a 3-nickel substituted ketone Eq. (28) [34]. This symmetrical coupling reaction is general for alkyl mercury compounds.

$$Ph \overset{O}{\diagdown} HgI \xrightarrow[\text{DMF}]{Ni(CO)_4} \left[Ph \overset{O}{\diagdown} Ni(CO)_3 I \right] \longrightarrow Ph \overset{O}{\diagdown} \overset{O}{\diagdown} Ph \qquad (28)$$

3.2.4 Carbonyl Addition

Addition of a homoenolate onto a carbonyl compound, which may be viewed as a "homo-aldol reaction", poses a significant conceptual problem. The nature of the homoenolate is such that the reaction with the internal carbonyl group is entropically greatly favored over that with the external target carbonyl group. Such a reaction can only compete with the internal electrophile, when it has a great enthalpic advantage. Up till now two types of homoenolates are known to undergo intermolecular addition onto carbonyl compounds, namely titanium [9, 10, 19] and zinc homoenolates of esters [33].

Trichlorotitanium homoenolate **2** smoothly adds to aldehydes at 0 °C [9, 10]. Due to the strongly acidic reaction conditions, however, addition products of aromatic aldehydes tend to undergo further transformations Eq. (29). The trichlorotitanium homoenolate does not react with ketones (Scheme 2).

$$Cl_3Ti \leftarrow O \overset{OR}{\diagdown} \xrightarrow[\text{CH}_2\text{Cl}_2]{\text{PhCHO}} Ph \overset{Cl}{\diagdown} COOR \qquad (29)$$

$$\underset{2}{} \qquad\qquad\qquad \underset{90\%}{}$$

Ligand exchange provided a simple and effective solution to these problems. Addition of 0.5 eq. of $Ti(O^iPr)_4$ to the trichlorotitanium homoenolate produces an alkoxytitanium species **19** which is more reactive than the original homoenolate, Eq. (30). This allows the addition to proceed under nearly neutral conditions. Preparation of 4-hydroxy esters by this method is summarized in Table 5. The structure of **19** has been studied by ^1H NMR spectroscopy [19].

$$Cl_3TiCH_2CH_2COOR \; + \; 1/2\; Ti(O^iPr)_4 \xrightarrow{CH_2Cl_2}$$

$$\underset{4}{}$$

$$Cl_2(O^iPr)TiCH_2CH_2COOR \; + \; 1/2\; TiCl_2(O\,Pr)_2$$

$$\underset{19}{} \qquad\qquad\qquad\qquad (30)$$

An even more reactive complex which is particularly useful for reactions with ketones is available by ligand exchange with $Ti(O^tBu)_4$, Eq. (31). The initial adduct tends to cyclize in situ to a γ-lactone, unless the reaction is carried out at low temperatures. The results are summarized in Table 6.

$$ 4 \ + \ 1/2 \ Ti(O^tBu)_4 \xrightarrow[\text{(2) } H_2O]{\text{1)}} \quad R^1 \underset{R^2}{\overset{OTi}{\diagdown}} COOR \longrightarrow \quad R^1 \underset{R^2}{\diagdown} \qquad (31) $$

These alkoxytitanium homoenolates show high propensity for equatorial attack in their ir reactions with substituted cyclohexanones (Table 6). The basic trend of their chemical behavior is similar to that of simple titanium alkyls [35]. Chemoselectivity of the reagent 19 is also noteworthy. The alkoxytitanium homoenolate reacts preferentially with an aldehyde even in the presence of a ketone Eq. (32). A notable difference of rate between the reaction with cyclohexanone and that with 2-methylcyclohexanone was also observed, the latter being far less reactive toward the homoenolate.

$$ PhCHO \ + \ Ph \overset{O}{\diagdown} \xrightarrow{\ 19\ } Ph \overset{OH}{\diagdown} COOR \ + \ Ph \overset{O}{\diagdown} \qquad (32) $$

97%   100%

Table 5. 4-Hydroxyesters by Homo-Reformatsky addition of isopropoxytitanium homoenolate (Ref. [19])

In line the with the chemistry of dialkylzinc [36], the zinc homoenolate is inert to carbonyl compounds in a variety of solvents, Eq. (33). Slow addition accurs only in an HMPA/THF mixture. When the reaction is conducted in halomethane in the presence of Me$_3$SiCl, however, a very rapid addition reaction occurs [33]. Control experiments indicate that the acceleration is due to the activation of the carbonyl group by Me$_3$SiCl. The activating effect of the chlorosilane disappears in ethereal solvents.

16

Table 6. Lactones by Homo-Reformatsky addition of *tert*-butoxytitanium homoenolate (Ref. [19])

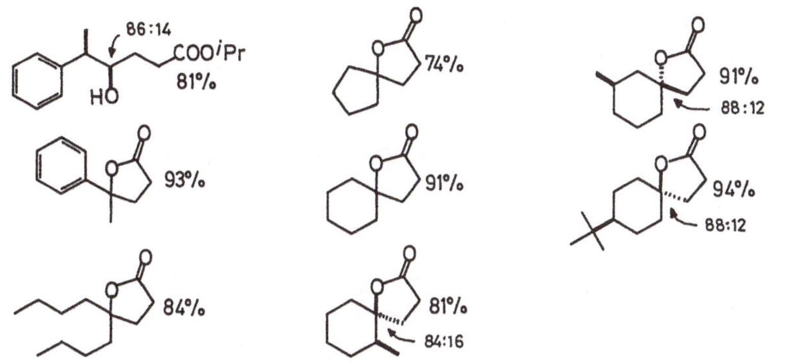

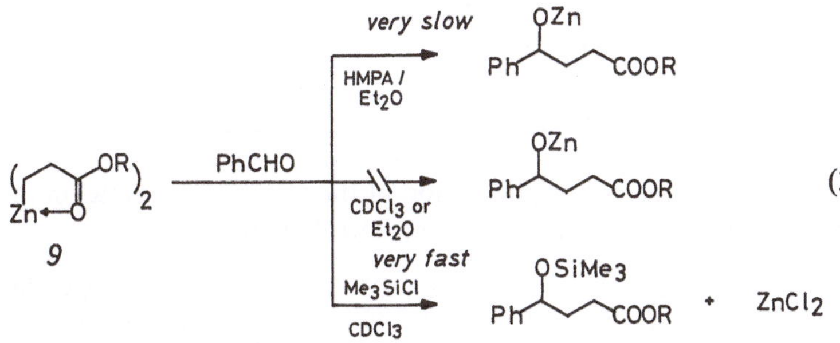

ZnCl$_2$ is generated in the Me$_3$SiCl mediated reaction, Eq. (33). Taking advantage of this fact as well as the accelerating effect of the chlorosilane, the cyclopropanes (**1, 6,** and **7**), and aldehydes have been coupled directly in the presence of ZnCl$_2$. ZnI$_2$, which in situ produces Me$_3$SiI, is an even more effective catalyst, only 1/1000 eq. of which suffices to obtain 4-siloxy esters in high yield Eq. (34) [33]. ZnI$_2$, but not ZnCl$_2$, promotes the addition reaction of 1-*tert*-butyldimethylsiloxy-1-isopropoxycyclopropane **20**. ZnI$_2$ also catalyzes the reaction of acetophenone or benzaldehyde dimethyl acetal, which fail to react under ZnCl$_2$ catalysis. Results of such of ZnX$_2$ catalyzed "homo-Reformatsky" reactions are summarized in Table 7.

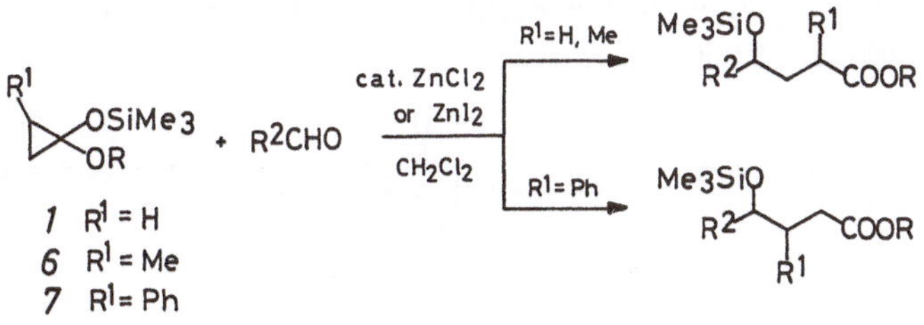

Table 7. 4-Siloxy esters by ZnX$_2$-catalyzed Homo-Reformatsky reaction (Ref. [29, 33])

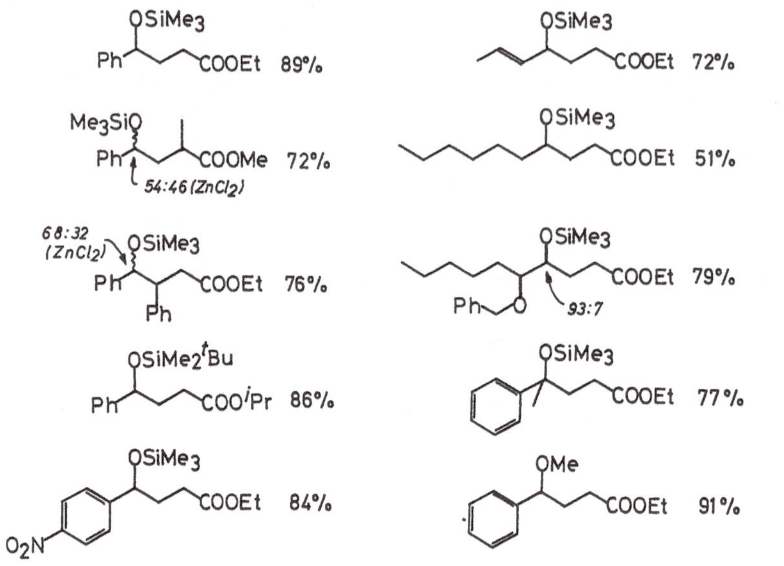

Interestingly, the carbonyl addition of the zinc homoenolate, in the presence of acid chloride, affords 4-acyloxyesters Eq. (35) [27].

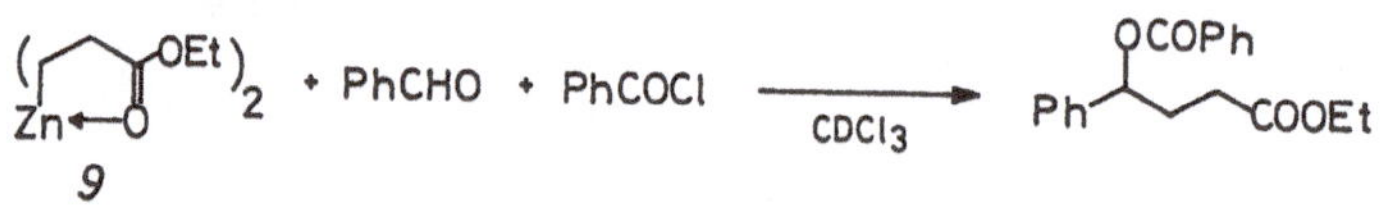

The fact that the above reactions allow isolation of 4-hydroxyesters, which are often unstable and lactonize quickly, is a merit of the homoenolate chemistry. Mesylation of the hydroxy group followed by appropriate operations provides stereocontrolled routes to γ-lactones and cyclopropane carboxylates [19]. Through application of such methodology steroid total synthesis has been achieved (Section 7).

3.2.5 Conjugate Additions

The mercurio ketone **4** upon reduction with NaBH$_4$ yields a radical which is trapped in situ by reactive acceptors such as vinyl ketones [37]. Treatment of a 1:3 mixture of mercurio ketone **4** and electron deficient terminal alkenes (or fumarate) in CH$_2$Cl$_2$ with a concentrated solution of NaBH$_4$ in water gives conjugate adducts, Eq. (36).

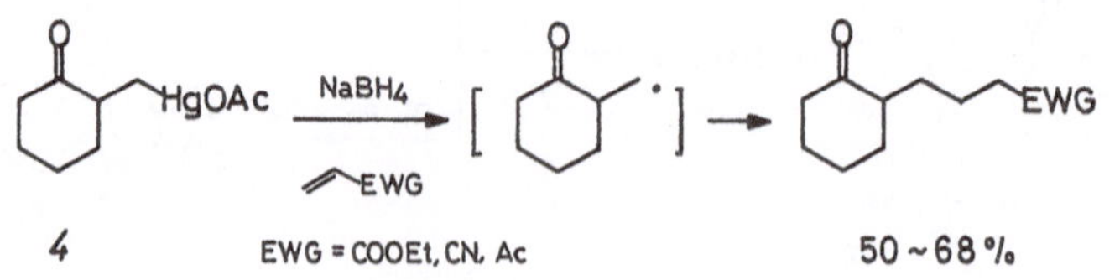

A highly intriguing isomerization of the "homoenolate radical" which probably proceeds via formation of a cyclopropyloxy radical, was noted in the reaction of a mercurio aldehyde [37]. The product ratio should reflect the ratio of internal/external trapping, and in fact the ratio of unrearranged and rearranged product depends on the concentration of the trapping reagent, Eq. (37).

$$
\begin{array}{cc}
0.33 \text{ M} & 18:82 \\
3.0 \text{ M} & 78:22
\end{array}
$$

Yield __%

(37)

When the stabilities of the two corresponding radicals are grossly different as in the case of gem-dimethyl homoenolate, one may obtain the product originating from the more stable homoenolate radical as the sole product, Eq. (38). A related isomerization reaction, albeit proceeding in the opposite direction, has often been recorded for anionic holmoenolates.

45%

(38)

A general method for the conjugate addition of ester homoenolate to an α,β-unsaturated ketone has been realized in a Cu(I)-catalyzed reaction of the zinc homoenolate [28]. Thus, successive addition of an enone, copper catalyst, and HMPA to the crude ethereal solution of the zinc homoenolate containing Me₃SiCl results in quantitative formation of the enol silyl ether (Eq. 39, Table 8). Me₃SiCl strongly accelerates the conjugate addition of intermediary organocopper species [38]. BF₃ · Et₂O also enhances the rate, but may give different stereochemical results [39].

$$
2 \text{ } \underset{\mathbf{1}}{\overset{OSiMe_3}{\triangleright\!\!<_{OR}}} + \text{ZnCl}_2 \xrightarrow{\text{Et}_2\text{O}} \text{Zn(CH}_2\text{CH}_2\text{COOR)}_2 + 2 \text{ Me}_3\text{SiCl}
$$

100%

(39)

The reaction with an acetylenic ketone produced an allenic enol ether (Table 8), which upon hydrolysis gave the corresponding enones as an E/Z mixture. Dimethyl acetylenedicarboxylate gave the (Z)-olefinic adduct as a major product in accordance with the general behavior of organocopper conjugate addition. A high degree of chemoselectivity has been observed in this conjugate addition.

Table 8. Products of conjugate addition of zinc homoenolate in the presence of Cu(I) and Me$_3$SiCl (Ref. [28, 29]

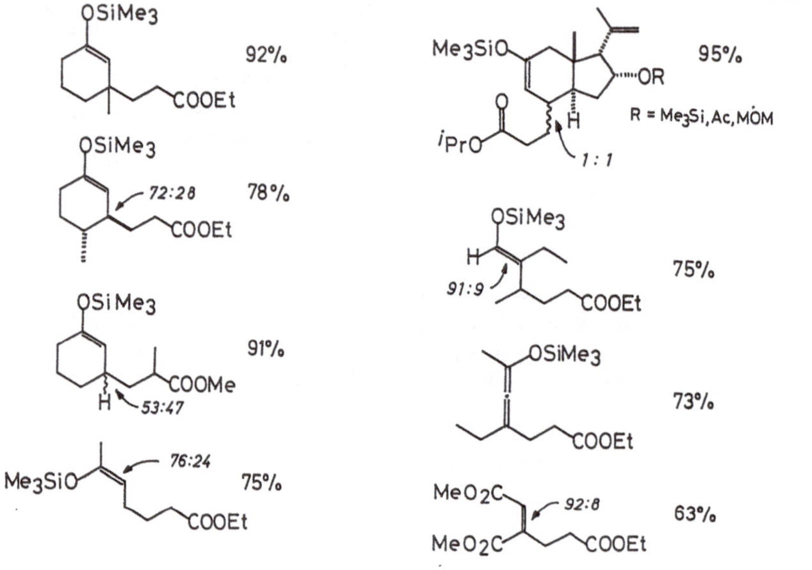

3.2.6 Substitution Reactions

Among the characterized metal homoenolates, only zinc homoenolate of alkyl propionate undergoes substitution reactions with electrophiles under suitable conditions. Two types of metal catalysts, copper(I) and metals of the nickel triad (e.g. Pd), have successfully been used to effect allylation, arylation, and vinylation reactions.

Zinc homoenolate reacts with allylic halides and diene monoepoxides under copper catalysis [29]. Treatment of the zinc nomoenolate with a catalytic amount of Cu(II) in a polar solvent (e.g. hexamethylphosphoramide, HMPA, N,N-dimethylacetamide, DMA) generates a copper species which undergoes clean S_N2' allylation reactions Eq. (40). Polar solvents not only accelerate the reaction but greatly improve the S_N2' selectivity. A variety of allylating reagents can be employed in this reaction (Table 9). The S_N2'/S_N2 ratio is particularly high (close to 100%) when the alkylated carbon bears no substituents. The reaction of *trans*-1,4-dibromo-2-cyclohexene gives a single stereoisomer which is presumably the *cis*-isomer arising from an *anti*-attack in the S_N2' reaction.

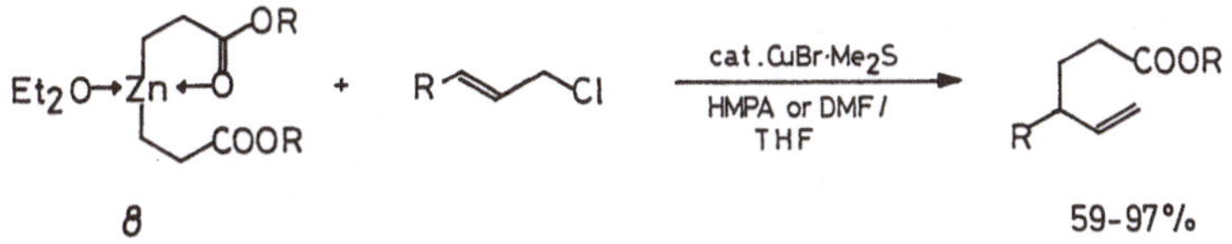

8

59-97%

(40)

The allylation and the conjugate addition (vide supra) of zinc homoenolate proceed under essentially the same conditions except that the latter requires the presence of Me_3SiCl as well. Due to this subtle difference, selective allyl substitution is possible even with an enone function is present in the same molecule (Table 9).

Table 9. Unsaturated esters by S_N2'-allylation of zinc homoenolate of esters (Ref. [29])

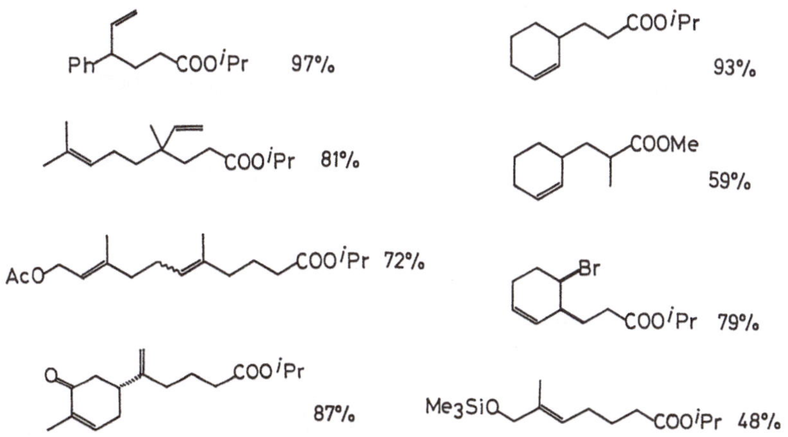

Copper catalyzed reaction of the homoenolate **9** with an acetal of an unsaturated aldehyde proceeds in an S_N2 manner, Eq. (41), the net reaction representing a 1,2-addition of the homoenolate. The presence of Me_3SiCl or $BF_3 \cdot Et_2O$ is mandatory in this reaction [29].

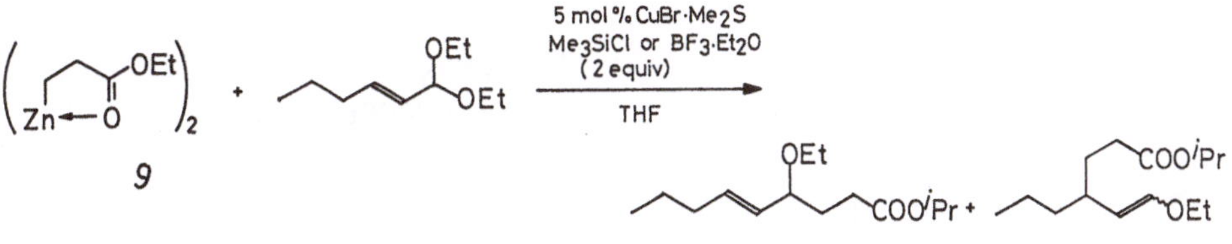

9

65% (94 : 6)   (41)

The nickel catalyzed reaction of **9** with an allyl chloride also proceeds in an S_N2 manner Eq. (42). Nickel dichloride complexed with a bidentate ligand (e.g. 1,2-bis(diphenylphosphino)ethane) effects a highly selective (>99%) S_N2 reaction with cinnamyl chloride [29].

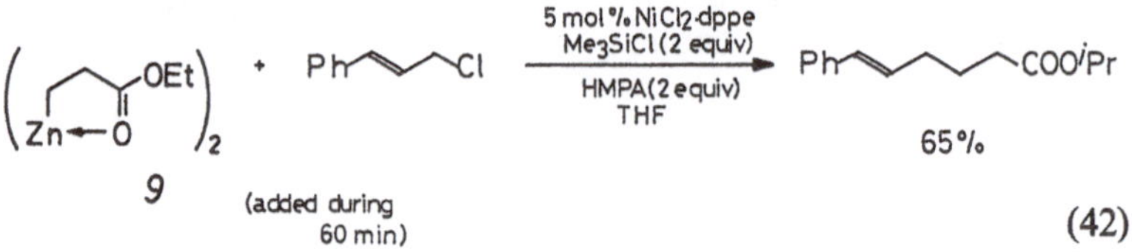

$$
\left(\underset{\underset{9}{Zn\leftarrow O}}{\overset{\displaystyle OEt}{\diagup}}\right)_2 + Ph\diagdown\diagup Cl \xrightarrow[\substack{HMPA(2\,equiv)\\THF}]{\substack{5\,mol\%\,NiCl_2\cdot dppe\\Me_3SiCl\,(2\,equiv)}} Ph\diagup\diagdown\diagup COO^iPr
$$

(added during 60 min)   65%   (42)

Substitution reactions involving aryl and vinyl halides have been achieved with the aid of palladium catalysts, among which $PdCl_2(o\text{-}Tol_3P)_2$ with $o\text{-}Tol_3P$ ligands is the most efficient [40]. A variety of aryl bromides and iodides can serve as the electrophile in this reaction (Table 10). The coupling with vinyl halides occurs stereoselectively. Vinyl trifluoromethanesulfonates (vinyltriflates), readily available from ketones [41], react rapidly with the homoenolate. Mild reaction conditions keep ketone functions intact during the arylation reaction.

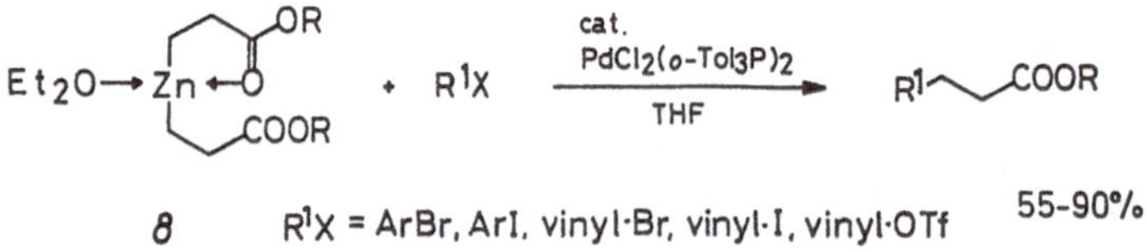

$$
Et_2O \rightarrow Zn \overset{\displaystyle OR}{\underset{\displaystyle COOR}{\diagdown}} + R^1X \xrightarrow[THF]{\substack{cat.\\PdCl_2(o\text{-}Tol_3P)_2}} R^1\diagup\diagdown COOR \qquad (43)
$$

8 $R^1X = ArBr, ArI, vinyl\text{-}Br, vinyl\text{-}I, vinyl\text{-}OTf$ 55-90%

Table 10. Arylated and vinylated propionates by palladium-catalyzed reaction of zinc homoenolate (Ref. [29, 40])

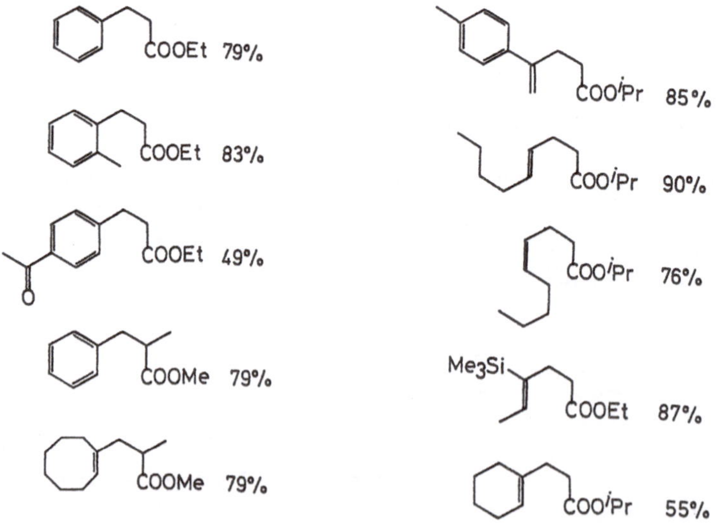

3.2.7 Acylation

The reaction of zinc homoenolate **9** with acid chlorides in ethereal solvents containing 2 equiv of HMPA rapidly produces 4-ketoesters in high yield Eq. (44) [33]. A palladium catalyst [40] (or less effectively a copper catalist) [28] accelerates the reaction. This is in contrast to the cyclopropane formation in a nonpolar solvent see (Eq. 22 above).

$$Et_2O \rightarrow Zn \longleftarrow O \quad + \quad R^1COCl \quad \xrightarrow[\substack{HMPA \text{ or } DMA/ \\ Et_2O}]{cat. PdCl_2(Ph_3P)_2} \quad R^1 \underset{O}{\overset{O}{\diagup}} COOR$$

50-93%

8 $R^1 = {}^tBu, \ PhCH_2CH_2, \ Ph, \ Me_2C{=}CH$ (44)

4 Other Routes to Reactive Metal Homoenolates

This section presents a summary of some recent reports on the chemistry of reactive metal homoenolates, which are generated by methods other than the siloxycyclopropane ring cleavage.

4.1 Anionically Protected Lithium Homoenolate

An early effort to generate a 3-lithiated propionic acid derivative and react it with (external) electrophiles was reported in 1978 [42]. Since simple 3-lithioesters failed to undergo the required reaction, the alkyl carboxylate portion was protected by preceding conversion to the carboxylate anion. Treatment of lithium 3-bromo-propionate with lithium naphthalide generated the desired dilithiated propionic acid, which gave moderate yields of γ-hydroxy acid addition products with carbonyl compounds, Eq. (45).

$$Br{\sim}COOH \quad \xrightarrow[\substack{2) \ Li \ Naphthalide}]{1) \ BuLi} \quad Li{\sim}COOLi \quad \xrightarrow{E^{\oplus}X^{\ominus}} \quad E{\sim}COOH$$

$$E^{\oplus}X^{\ominus} = \underset{R^2}{\overset{R^1}{\diagup}}{=}O$$

(45)

Treatment of 3-stannylpropionamide with two equivalents of butyllithium at low temperature (−78 °C) generates a similar dianionic homoenolate, which reacts with standard electrophiles Eq. (46) [43]. It is interesting to note that the propionate moiety rather than the butyl groups is selectively cleaved off the tin in the tin-lithium exchange.

$$Bu_3Sn{\sim}CONHPh \quad \xrightarrow[DABCO/THF]{2 \ BuLi} \quad Li{\sim}\underset{Li}{CONPh} \quad \xrightarrow{E^{\oplus}X^{\ominus}} \quad E{\sim}CONHPh$$

$$E^{\oplus}X^{\ominus} = Me_3SiCl, \ RBr, \ \underset{R^2}{\overset{R^1}{\diagup}}{=}O$$

(46)

Dianions of the above types may not fall into the category of homoenolate in a strictly formal sense. Nevertheless the amide dianion does show a behavior typical of the homoenolate. Thus, the reaction of the isotopically labeled stannylpropionate results in scrambling of the label probably via a cyclopropane intermediate Eq. (47) [44]. As the result of such an equilibration, isomerization of α-methyl and α-phenyl substituted propionate homoenolates may occur to give the thermodynamically more favorable isomers, respectively.

$$\text{Bu}_3\text{Sn}\overset{\text{CONHPh}}{\underset{\text{D D}}{\diagdown}} \xrightarrow[\text{2) } -40°C]{\text{1) } -70°C} \left[\text{Li}\overset{\text{CONPh}}{\underset{\text{D D}}{\diagdown}}_\text{Li} \rightleftharpoons \overset{\text{LiO} \underset{\text{Li}}{\text{NPh}}}{\underset{\text{D}}{\triangle}} \rightleftharpoons \text{Li}\overset{\text{D D}}{\underset{\text{Li}}{\diagdown}}\text{CONPh} \right] \quad (47)$$

4.2 From 3-Stannyl- to 3-Titaniopropionate

Two groups independently reported the formation of titanium homoenolates by the transmetalation reaction of 3-stannyl-propionate esters with TiCl$_4$, Eq. (48) [45, 46]. Amide homoenolates become available along this route [47]. The trichlorotitanium species thus obtained have been shown (^{1}H NMR) to be similar to that generated along the siloxycyclopropane route and indeed exhibit very similar reactivities. This method does provide a conventient alternative to the siloxycyclopropane route.

$$\text{Bu}_3\text{Sn}\overset{\text{COX}}{\diagdown} + \text{TiCl}_4 \longrightarrow \underset{\text{Cl}_3\overset{}{\text{Ti}}\leftarrow\text{O}}{\overset{X}{\diagdown}} \xrightarrow[50-85\%]{\text{RCHO}} R\overset{\text{OH}}{\diagdown}\overset{}{\diagdown}\text{COX} \quad (48)$$

$$X = OR, NR_2$$

Isotopic labelling, Eq. (49), indicated that the transmetalation occurs via direct tin-carbon bond cleavage rather than via intermediate formation of an alkoxy-cyclopropane [45].

$$\text{Bu}_3\text{Sn}\overset{\text{COOMe}}{\underset{\text{D D}}{\diagdown}} \begin{array}{c} \xrightarrow{-/\vdash} \overset{\text{MeO OTi}}{\underset{\text{D}}{\triangle}}_\text{D} \longrightarrow \underset{\text{Cl}_3\overset{}{\text{Ti}}\leftarrow\text{O}}{\overset{\text{D}_2}{\diagdown}}\text{OMe} \\ \xrightarrow[\textit{direct exchange}]{} \underset{\text{Cl}_3\overset{}{\text{Ti}}\leftarrow\text{O}}{\overset{\text{D D}}{\diagdown}}\text{OMe} \end{array} \quad (49)$$

4.3 From 3-Stannyl- to 3-Palladioketoxime

Selective transfer of the homoenolate moiety characterizes the transmetalation from tin to titanium described in the previous section. A very facile transmetalation of the same kind occurs for 3-stannylketoxime, Eq. (50) [48]. Reaction of the (E)-3-(tributyltin)ketoxime with dichlorobis(benzonitrile)palladium in CH$_2$Cl$_2$ at 0 °C for 30 min yields 78% of the transmetalation product. The facility with which this reaction produced the stable palladium complex, and the high selectivity with which the oxime group carrying alkyl group is cleavech off, are remarkable. The reaction certainly proceeds with prior coordination of the metal to the oxime group, but the precise mechanism of the transmetalation remains unclear.

$$\text{Bu}_3\text{Sn} \diagdown \overset{\text{NOH}}{\diagup} \quad + \quad \text{PdCl}_2 \quad \longrightarrow \quad \text{Cl}\overset{|}{\text{Pd}} \leftarrow \overset{\text{NOH}}{\diagup} \qquad (50)$$

4.4 From 3-Iodoesters to Alkyl β-Zincpropionate

A very practical route to zinc homoenolate involves reduction of 3-iodoesters with zinc/copper couple in the presence of a polar solvent, e.g. DMF, DMA [49] Eq. (51). The nature of the species obtained in this approach is not well-defined, but appears to be essentially the same as the one obtained along the siloxycyclopropane route. Acylation, arylation, and vinylation reactions have been reported.

$$\text{I} \diagdown \diagup \text{COOEt} \quad \xrightarrow[\substack{\text{DMA}/ \\ \text{C}_6\text{H}_6}]{\text{Zn/Cu}} \quad [\ \text{Zn(II)} \diagdown \diagup \text{COOEt}\ ] \quad \xrightarrow{\text{E}^+} \quad \text{E} \diagdown \diagup \text{COOEt}$$

$$(51)$$

5 Reactions of Transient Homoenolates Generated from Siloxycyclopropanes

There have been several reports on reactions of siloxycyclopropanes involving transient metal homoenolates which eluded characterization owing to their instability.

5.1 Iron(III) Homoenolates

Most ring cleavages of siloxycyclopropanes occur at the less substituted bond next to the siloxy group to give the more stable anionic species (Cf. Eq. 34). Chlorination with $FeCl_3$ in DMF provides a unique way to cleave the ring in the opposite direction, Eq. (52) [50]. Treatment of 1-siloxybicyclo[n.1.0]alkanes with $FeCl_3$ in DMF produces 3-chloroketones with cleavage of the internal bridging bond. Such regiochemistry has not been observed for ionic cleavages of siloxycyclopropanes, and must be due to the radical character of the intermediate formed in the reaction. It is not even clear whether a metal homoenolate is involved in this reaction at all, and the reaction is believed to proceed with initial cyclopropoxy radical formation. The regiochemistry is consistent with the results obtained for reactions of discrete "homoenolate radicals" (vide supra) for which thermodynamically more stable secondary radicals result by very rapid equilibration via cyclopropoxy radicals.

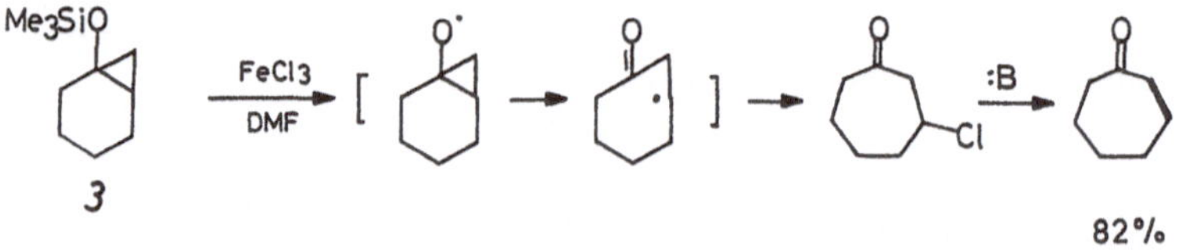

(52)

The overall sequence of cyclopropanation of a cyclic silyl enol ether, chlorination with FeCl$_3$, and dehydrochlorination represents a very reliable one-carbon ring expansion method for cycloalkanomer (Table 11).

Application of this ring enlargement and oxidation procedure to 2-alkoxy-1-siloxycyclopropanes incorporated in a large ring provides macrocyclic diketones which act as potent chelating agents for metal cations [50].

Table 11. Synthesis of 2-cycloalkenones by FeCl$_3$-oxidation (Ref. [50])

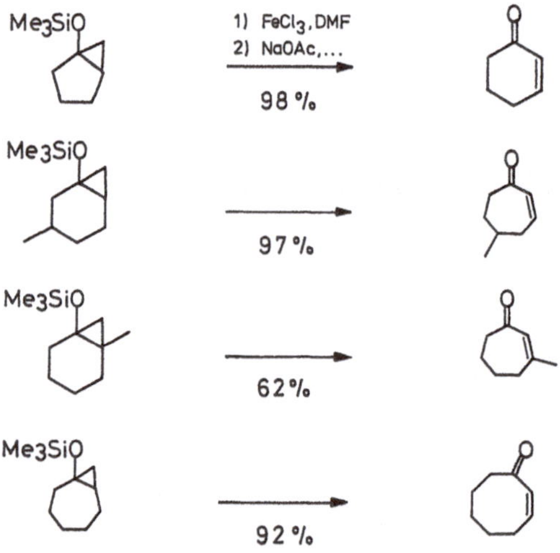

5.2 Copper(II) Homoenolates

Reaction between a siloxycyclopropane and Cu(BF$_3$)$_2$ in ether gives a product due to symmetrical coupling of two homoenolate moieties (Eq. 53, Table 12) [51]. This is particularly noteworthy as a simple route to 1,6-ketones superior to classical approaches such as the Kolbe electrolysis [52]. Several lines of evidence suggest the intermediacy of Cu(II) homoenolates. AgBF$_3$ and CuF$_2$ effect the same reaction albeit with lower yields. The reactions with cupric halides give complex mixtures containing these homocoupled product and halogenated compounds [53].

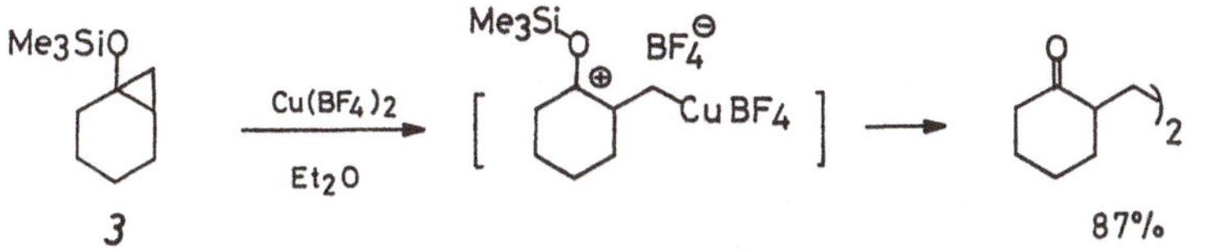

(53)

Table 12. 1,6-Diketones by homocoupling of siloxycyclopropanes in the presence of AgBF$_4$ or Cu(BF$_4$)$_2$ (Ref. [5])

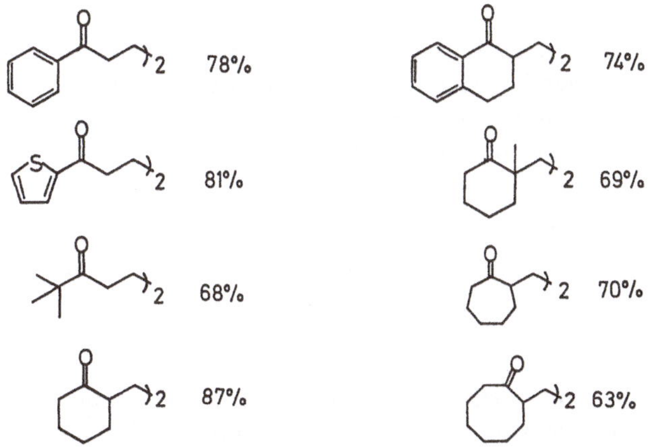

When the siloxycyclopropane and Cu(II) are allowed to react in the presence of dimethyl acetylendicarboxylate or propynyl phenyl sulfone in wet CH$_2$Cl$_2$, conjugate adducts are obtained in good yields Eq. (54) [54]. The same reaction conducted in CH$_2$Cl$_2$ containing D$_2$O gives the stereoselectively α-deuterated products, indicating the intermediacy of vinylcopper intermediates. Under strictly dry conditions homocoupling products are produced from these vinyl copper species.

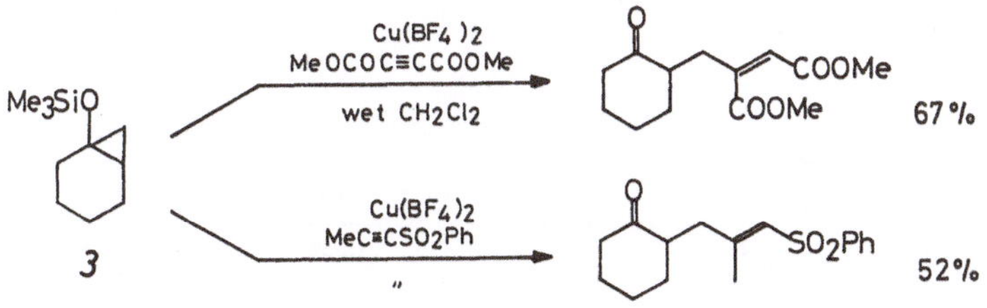

(54)

27

5.3 Silver Homoenolates

A silocyclopropane reacts with an allylic chloride in the presence of AgF in aqueous ethanol to give an allylated product Eq. (55) [55]. In this case, the structure of the product does not reflect the regiochemistry of the nucleophilic attack on the starting allyl halide.

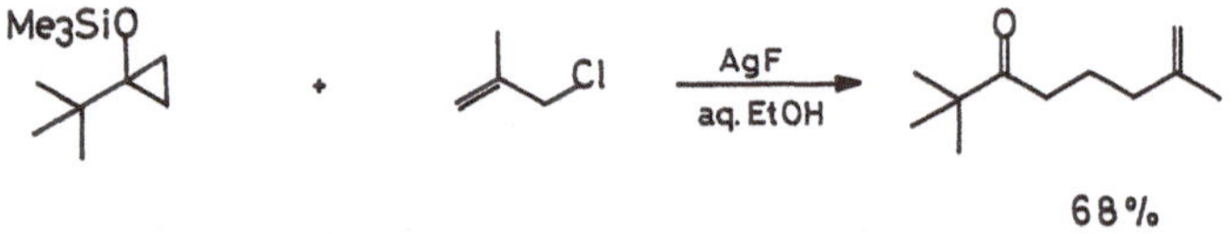

$$\text{(image of reaction)} \tag{55}$$

68%

5.4 Palladium Homoenolates

Polladium(II) chloride or its phosphine complex smoothly reacts with siloxy-cyclopropane 1 to produce acrylic ester and a palladium mirror. This reaction probably involves the formation of a chloropalladium homoenolate followed by elimination of palladium hydridochloride (Eq. (56) [56].

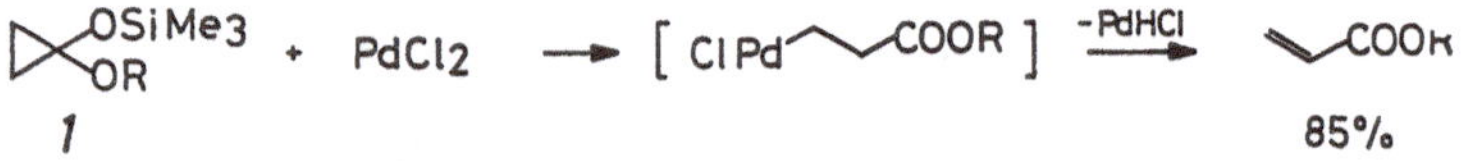

$$\text{(image of reaction)} \tag{56}$$

85%

Nevertheless, the palladium homoenolate can serve as a useful reaction intermediate. If certain requirements are met, such an organopalladium homo-enolate can be generated directly from the siloxycyclopropane, and then allowed to reductively eliminate to give an organylated homoenolate. Heating a mixture of siloxycyclopropane 1 and an aryl triflate in benzene in the presence of a palladium catalyst effects such a transformation to give 3-arylpropionate Eq. (57) [57]. A siloxycyclopropane corresponding to a ketone homoenolate (e.g. 3) also undergoes reactions with aryl triflates, now in HMPA, to give arylated products Eq. (58) [57]. These results are summarized in Table 13.

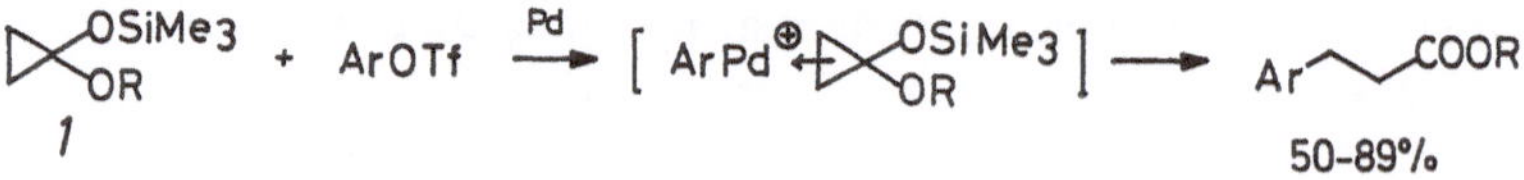

$$\text{(image of reaction)} \tag{57}$$

50-89%

$$\text{(image of reaction)} \tag{58}$$

50-70%

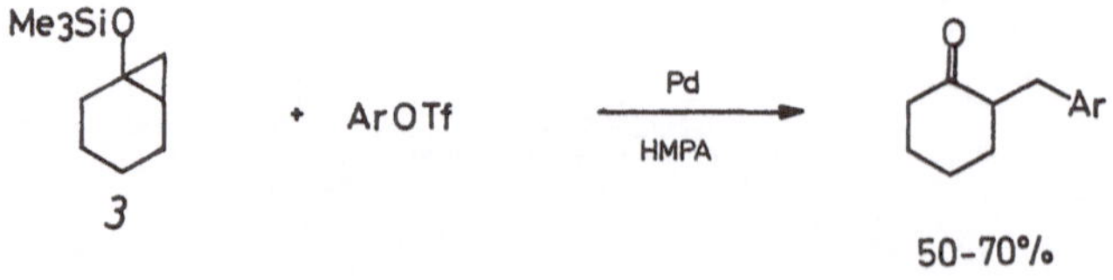

The arylation reaction is not applicable to aryl bromides and iodides. Another line of experimental evidence also indicates that an arylpalladium *halide* complex, instead of an arylpalladium *triflate* complex, is not electrophilic (Lewis acidic) enough to cleave the cyclopropane ring.

Table 13. Arylated esters and ketones by direct arylation of siloxycyclopropanes (Ref. [57])

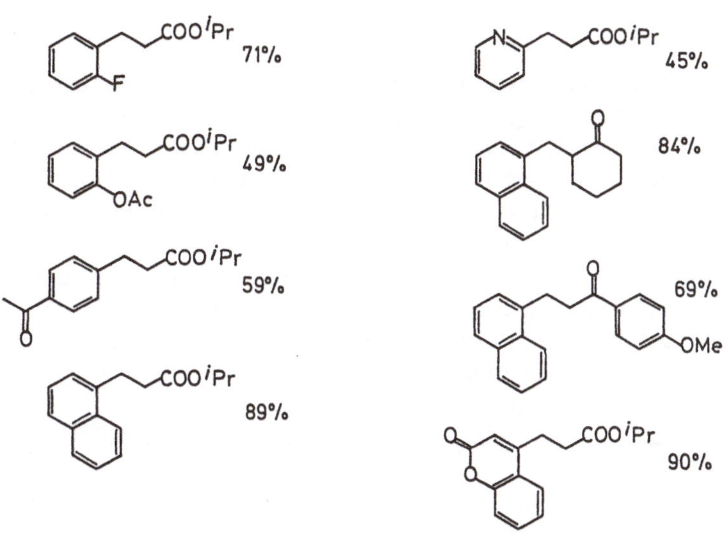

Reaction of siloxycyclopropane **1** with acid chlorides in the presence of a palladium catalyst also proceeds cleanly to give 4-ketoesters in high yields (Eq. 59, Table 14) [57]. Chloroform is a suitable solvent. Kinetic studies have revealed that the interaction between **1** and an acylpalladium chloride complex is the rate limiting step.

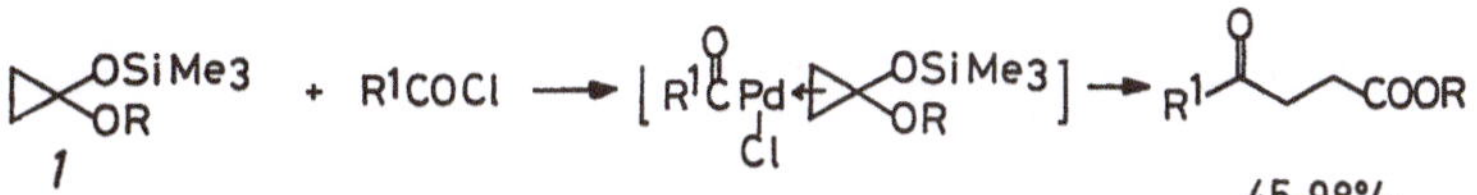

$$45\text{-}98\% \tag{59}$$

Table 14. Keto esters by direct palladium catalyzed acylation of siloxycyclopropanes (Ref. [56])

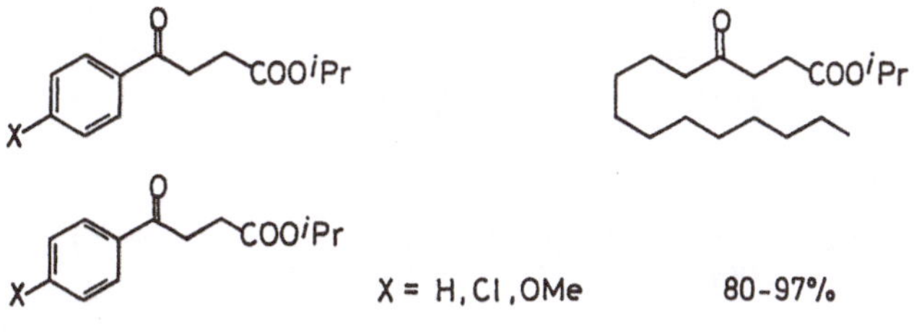

X = H, Cl, OMe   80-97%

If the cyclopropane is heated in $CHCl_3$ in the presence of a catalytic amount of palladium-phosphine complex under CO atmosphere, a symmetrical coupling product incorporating one molecule of CO forms in high yield, Eq. (60) [57].

$$60\text{-}79\% \quad (60)$$

5.5 Zinc Homoenolates

In view of the remarkable stability of metal homoenolates of esters, the existence of homoenolate species containing a 3-halo substituent, i.e. zinc carbenoid moiety connected to an ester group, appeared to be possible. Indeed, when a silyl ketene acetal is treated with a carbenoid generated from $CHBr_3$ and Et_2Zn, two types of highly intriguing reactions ensue [58]. With a purely aliphatic substrate, Eq. (61), an alkyl cyclopropylcarboxylate due to intramolecular β-CH-insertion of the intermediate zinc carbenoid formed. When the substrate contained an olefinic double bond in the vicinity of the carbenoid function, Eq. (62), in particular an intermediate derived from an α,β-unsaturated ester, internal cyclopropanation occurred to give bicyclic or tricyclic product (Table 15).

$$(61)$$

$$(62)$$

Table 15. Alkyl cyclopropanecarboxylates by cabene addition onto ketene acetals (Ref. [58])

5.6 Lead(IV) Homoenolate

In contrast to the other examples in this review, the reaction of 1-siloxybicyclo[n.1.0]alkenes with Pb(OAc)$_4$ in acetic acid somewhat surprisingly leads to cleavage of two bonds to yield alkenoic acids, Eq. (63) [59]. The reaction involves the fission of both bonds connected to the carbon bearing the siloxy group. This reaction is considered to involve primary in situ conversion of the siloxycyclopropane to the corresponding cyclopropanol Indeed, the siloxycyclopropane **3** reacts with Pb(OAc)$_4$ in dichloromethane to give the β-acetoxyketone, Eq. (64), while the corresponding cyclopropanol undergoes the same fragmentation as with Pb(OAc)$_4$ in acetic acid.

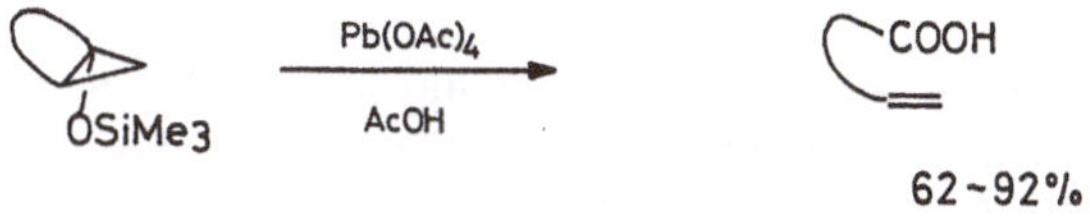

$$\text{(63)}$$

$$62\text{–}92\%$$

$$\text{(64)}$$

3

$$75\%$$

The observed stereoselectivity of this ring cleavage reaction [60] favors a mechanism, Eq. (65), in which formation of a lead(IV) homoenolate by cleavage of the external bond with inversion of configuration is followed by rapid cleavage of the second bond.

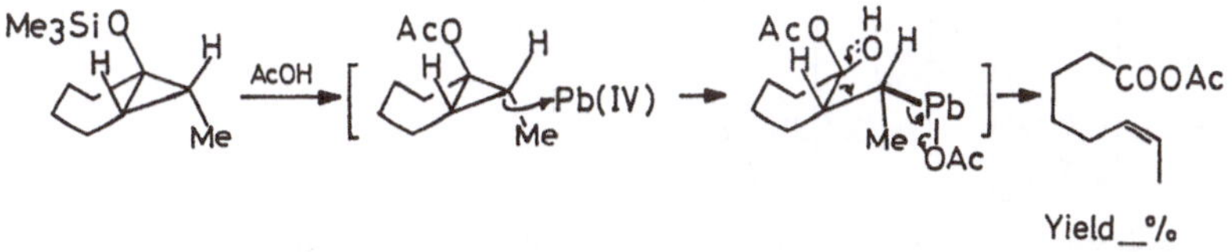

$$\text{(65)}$$

Yield__%

6 Mechanism of Cyclopropane Ring Cleavage

Except for some of the above reactions which proceed via radical intermediates, many of the ring cleavage reactions are likely to involve initial interaction of the metal and the cyclopropanering. The nature of this interaction between 1-siloxy-1-alkoxycyclopropane **1** with metals has been investigated by the authors in some detail, and the results are summarized below.

The Dewar-Chatt-Duncanson model of the binding of an olefin in a transition metal complex involves two types of interactions. Transfer of electron density from the relatively high-lying olefinic π-orbital to the metal (cf. 20) represents a Lewis acid Lewis base interaction (σ-bonding). A metal-olefin π-bond due to interaction of a filled metal orbital with the π^*-orbital of the olefin (back donation) constitutes another mode of bonding (cf. **21**).

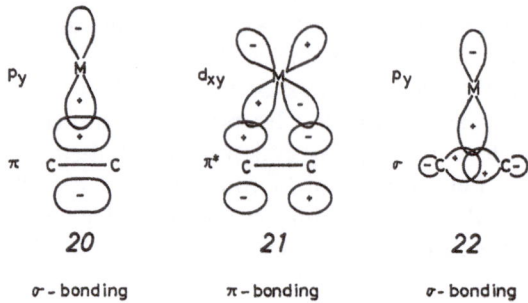

20	21	22
σ - bonding	π - bonding	σ - bonding

Organometallic bonding to σ bonds possessing high p-character, in contrast, is due mainly to the electron donation from the carbon-carbon σ-bond into an empty metal orbital (cf. **22**), because the antibonding σ^*-orbital is generally inaccessible for effective back donation from the metal [61]. It follows from such considerations that the donor character of the σ-bond of a cyclopropane derivative will be decisive. In addition, the Lewis acidity of the metal will be more important for the binding with cyclopropane. The summary chart of metals which cleave the alkoxysiloxycyclopropane **1** (Table 16) [11] offers an idea for the kind of Lewis acidity needed.

Table 16. List of metals, chlorides of 1 which react with cyclopropane *1* to form metal homo-enolate (Ref. [11])[a]

1	2	3	5	12	13	14	15	16
~~Li~~					B^I			
~~Na~~	~~Mg~~				Al^{III}	Si^{IV}		
~~K~~	~~Ca~~	Ti^{IV} ~~Ti^{III}~~		Zn^{II}	Ga^{III}	Ge^{IV}		
~~Rb~~	~~Sr~~	Zr^{IV}	Nb^V	Cd^I		$Sn^{IV}Sn^{II}$	Sb^V	Te^{IV}
~~Cs~~	~~Ba~~		~~Ta^V~~	Hg^I		~~Pb^{II}~~ Bi^{III}		

[a] Signs X and / indicate no reaction and formation of no detectable homoenolate, respectively

There are a few points to be addressed in order to understand the mechanism of homoenolate formation. Several lines of experimental evidence for the reaction of 1-alkoxy-1-siloxycyclopropanes have provided insights into the nature of the metal interacting with the siloxycyclopropane.

The initial interaction of the metal with the substrate (Scheme 4) may occur either on the siloxy oxygen to form a metal alkoxide (path (a)) or directly on the strained sigma bond to form an intermediate containing an oxygen-stabilized cationic center (path (b)). Molecular orbital considerations indicate that an orbital controlled interaction with a Lewis acid should occur preferentially at the ring carbons, which have the larger HOMO coefficient. Studies of the exchange of a Me_3Si group between Me_3SiOTf and silyl ethers eliminate the possibility of path (a) for a number of metal halides, including $ZnCl_2$ and $SnCl_4$ [11].

Scheme 4

Some experimental data indicate the formation of a cationic species as postulated for path (b) and in turn show that the Si-O bond remains intact in the transition state. Evidence comes from the comparison of the reactivities of 1-trimethylsiloxy- (1) and 1-*tert*-butyldimethylsiloxycyclopropanes (23) in some reactions.

The reaction of 1-isopropoxy-1-trimethylsiloxycyclopropane (1, R = i-Pr) with TiCl$_4$ yields the titanium homoenolate of isopropyl propionate 2 (R = i-Pr) and Me$_3$SiCl (Eq. see 9). That of 23, in contrast, yields a considerable fraction of the homoenolate of silyl propionate as well, Eq. (66) [19]. Apparently, the loss of an isopropyl group from an oxygen-stabilized cationic intermediate competes with that of the silyl group with bulky substituents.

Kinetic analysis of the palladium catalyzed acylation reaction of 1 (R = i-Pr) and 23 indicates that the rate does not depend on the bulk of the trialkylsilyl substituent. Since the rate limiting step of this reaction is the interaction of a coordinatively unsaturated acylpalladium chloride with the cyclopropane (Cf. Eq. 59), the observed independence can reasonably be taken as an evidence that the Si—O bond remains intact in the transition state [56]. Semiquantitative data on the cleavage of 1 (R = i-Pr) and 23 with ZnCl$_2$ in ether, Eq. (13), led to the same conclusion [27].

The mechanisms of the ring cleavages which lead to Ti, Zn, and Pd homoenolates are similar to each other as described above in that the oxygen-stabilized cation is involved (path (b) in scheme 4). The fact that the regiochemistry of the ring cleavage strongly depends on the nature of the forming carbon-metal bond (Cf. Eq. 67), however, indicates that they are not identical.

(67)

Scheme 5 summarizes the regiochemistry of ring opening of methyl- (**6**) and phenyl-substituted alkoxy-siloxy-cyclopropane (**7**) by LiOMe, ZnCl$_2$, HgCl$_2$, and TiCl$_4$, as well as the ^{13}C NMR chemical shifts of the respective metal methyls [27]. The NMR data correlate with the nature of the metal-carbon bond: The more polarized it is, the less positive the chemical shift of the methyl group.

M$^+$	path(a) : path(b)		^{13}C NMR (ppm)
	R^1 = Me	R^1 = Ph	CH$_3$M
MeO*Li*/MeOD	100 : 0	0 : 100	<−10
ZnCl$_2$/PhCHO	>98 : 2	2 : >98	~−4
Hg(OAc)$_2$/EtOH	>95 : 5	22 : 78	1~10
*Ti*Cl$_4$/PhCHO	61 : 39	22 : 78	~40

Scheme 5

The regiochemistry with lithium methoxide and zinc chloride clearly reflects the stability of the incipient formal anionic charge on the carbon. Such a trend is less obvious for those metals which form covalent bonds to carbon. Electronic and steric factors compete with each other in the cleavage of the phenylsubstituted alkoxy-siloxy-cyclopropane **7**, in particular. In conclusion, the transition state (**24**) appears to anticipate the nature of the intermediate **25**, which on one hand reflects the property of R^1 and on the other that of the metal carbon bond (Eq. 67). More detailed information on the type of ring cleavage, namely, the choice between the side-on (i.e. **24**) or the end-on interaction (attack of the metal at the back-lobe of the opening C—C bond), is currently lacking: The mode of the mercury induced cleavage of cyclopropanols is sensitive to ring substituents [7].

7 Homoenolates in Natural Product Synthesis

As demonstrated in the preceding chapters, metal homoenolate chemistry centered around the ring cleavage of siloxycyclopropanes has realized a number of straightfoward synthetic transformations only some of which habe been possible with other existing methods.

The current state of metal homoenolate chemistry is still in its infancy, and very few examples of the actual applications have been reported. However, the diversity of reactivities of the homoenolates shown in the preceding paragraphs will definitely lead future activities towards fruitful applications.

7.1 Steroidal Side Chains:
Depresosterol and 24-epi-Depresosterol [63]

Stereocontrolled functionalization of steroidal side chains in nature is closely related to the function of steroids in living organisms. The regio- and stereochemistry of functional groups exert strong influence on the biological activities of steroids. Due to their multi-functional nature, the homoenolates provide an effective tool for synthetic efforts in this field.

Depresosterol (26) is a marine natural product isolated from a soft coral in the Red Sea [64]. The structure of this sterol, in particular, the configuration at C^{20} and C^{22}, was assigned by comparative NMR studies, and that at C^{24} was left unassigned. The latter configuration was of interest with respect to the biological significance of the sterol. The density of hydroxyl groups is higher than in any of the known sterols, the substitution pattern being reminiscent of the insect molting hormone, ecdysone. The synthesis of depresosterol 26 demonstrates the potential of the homoenolate as a multi-functional three carbon building block, which corresponds to the C_3 fragment C^{23}, C^{24}, C^{25} in 26. The stereorational synthesis of both epimers at C^{24} finalized the structure determination of this sterol [63].

26

The synthesis started with the "homo-Reformatsky" reaction between the alkoxytitanium homoenolate Eq. (30) and aldehyde 27, which afforded the product 28 with Cram orientation with > 6:1 selectivity. An inversion at the sterically hindered C^{22} position was readily achieved by mesylation, followed by KOH treatment in hot aqueous MeOH to give lactone 30 after acid-catalyzed lactonization.

Synthesis of the 24R isomer was commenced by stereoselective hydroxymethylation of the enolate of lactone 30. Introduction of methyl groups at C^{25} and C^{26} was achieved by addition of MeLi to give 24R depresosterol (33). Alternatively, trapping of the lactone enolate with acetone followed by $LiAlH_4$ reduction gave the 24S epimer (34). Spectral comparison indicated that the 24R sterol is identical with the natural product.

This approach not only enabled the preparation of a simpler structure such as that of ecdysons (cf. 31), but also provides ways to introduce functional groups at C^{23} and C^{24} positions of the side chain. As an example, treatment of the mesylate 29 with non-nucleophilic tert-butoxide gives the cyclopropane 32 which

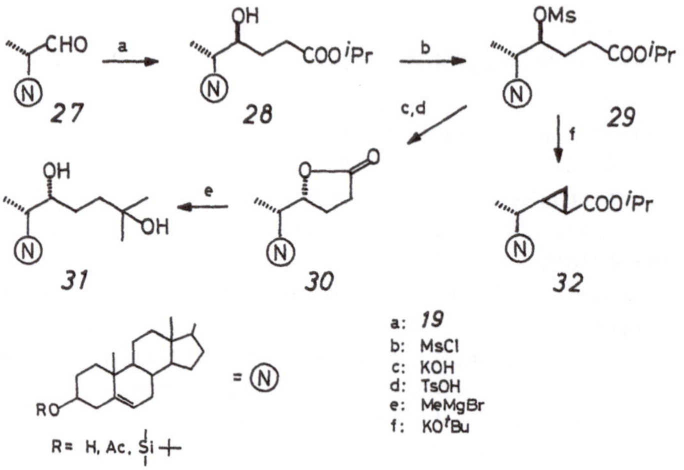

Scheme 6

can serve as a precursor to the structurally intriguing demethylgorgosterol [63].

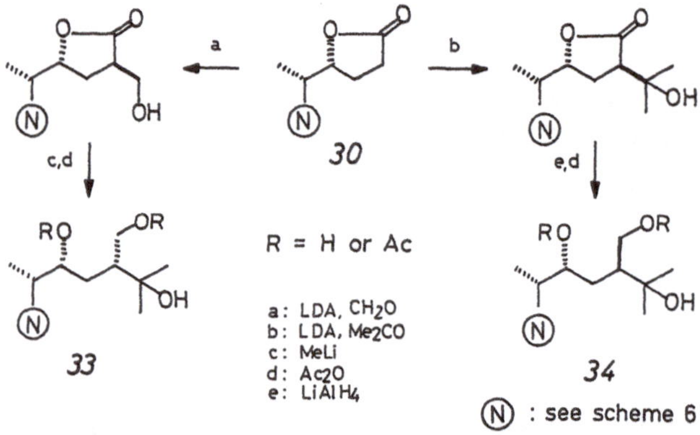

Schemes 6, 7

7.2 (±)-Cortisone [39]

Conjugate addition of catalytically generated zinc homoenolate (cf. Eq. 39) has been used for the stereoselective synthesis of a key intermediate to (±)-cortisone (**38**). Precedents of the cuprate addition onto a *trans*-fused hydrindane such as **35** was expected to give **36** but attack of the nucleophilic copper reagent from the desirable, less hindered α-side. Although treatment of the enone **35** with the zinc homoenolate under standard condition (Eq. 39, $Me_3SiCl/HMPA/$ cat. $CuBr \cdot Me_2S$) afforded the adduct in quantitative yield, the product was a 1:1 mixture of **36** and **37**. In contrast, use of $BF_3 \cdot Et_2O$ instead of Me_3SiCl gave the desired isomer **36** with better than 95% selectivity; with some additional steps the total synthesis of **38** was completed. The precise reason for the change of the stereochemistry is currently unclear.

Scheme 8

7.3 (±)-Muscone [20]

The acyloin condensation of α,ω-diesters in the presence of Me₃SiCl represents a highly efficient method for the synthesis of large rings. The application of the convenient sequence of cyclopropanation/oxidative ring cleavage for ring expansion cf. Eq. (52) to the appropriate macrocyclic enediol silyl ether **39** provided, a short, high-yield (23% overall) synthesis of muscone (**40**) [20].

a: CH₂I₂
b: FeCl₃, DMF
c: isopropenyl acetate
d: Me₂CuLi
e: H₂, Pd/C

Scheme 9

7.4 (±)-Sabinene [58]

The remarkably efficient cyclopropanation reaction with an in-situ-generated zinc carbenoid, Eq. (61) provided a straightforward synthesis of the monoterpene sabinene (**42**) starting from the β-keto ester **41** [58].

Scheme 10

8 References

1. Nickon A, Lambert JL (1962) J. Am. Chem. Soc. 84: 4604
2. Werstiuk NH (1983) Tetrahedron 39: 205; Ryu I, Sonoda N. (1985) J. Syn. Org. Chem., Jpn. 43: 112; Nakamura E (1989) J. Syn. Org. Chem. Jpn. 47: 931
3. Seebach D (1979) Angew. Chem. Int. Ed. Engl. 18: 239; Evans DA, Andrews GC (1974) Acc. Chem. Res. 7: 147
4. Brewis S, Hughes PR: J. Chem. Soc., Chem. Commun. 1967: 71
5. Bennett MA, Watt R: J. Chem. Soc., Chem. Commun. 1971: 95; Bennett MA, Robertson GB, Watt R, Whimp PO: J. Chem. Soc., Chem. Commun 1971: 752
6. Hoppe D (1984) Angew. Chem. Int. Ed. Engl. 23: 932; see also ref. 8 in Ref. 10
7. DeBoey A, DePuy CH (1970) J. Am. Chem. Soc. 92: 4008; Gibson DH, DePuy CH (1974) Chem. Rev. 74: 605
8. Ryu I, Matsumoto K, Ando M, Murai S, Sonoda N (1980) Tetrahedron Lett. 21: 4283
9. Nakamura E, Kuwajima I (1977) J. Am. Chem. Soc. 99: 7360
10. Nakamura E, Kuwajima I (1983) J. Am. Chem. Soc. 105: 651
11. Nakamura E, Shimada J, Kuwajima I (1985) Organometallics 4: 641
12. Denis JM, Conia JM: Tetrahedron Lett. 1972: 4593, Conia JM, Girard C: Tetrahedron Lett. 1973: 2767: Girard C, Conia JM: Tetrahedron Lett. 1974: 3327
13. le Goaller R, Pierre JL: Bull. Chim. Soc. Fr. 1973: 153
14. Rubottom GM, Lopez MI (1973) J. Org. Chem. 38: 2097
15. Murai S, Aya T, Sonoda N (1973) J. Org. Chem. 38: 4354
16. Rousseaux G, Slougui N (1983) Tetrahedron Lett. 24: 1251
17. Furukawa J, Kawabata N, Nishimura J: Tetrahedron Lett. 1966: 3353
18. Miyano S, Izumi Y, Fujii H, Hashimoto H: Synthesis 1977: 700
19. Nakamura E, Oshino H, Kuwajima I (1986) J. Am. Chem. Soc. 108: 3745
20. Ito Y, Saegusa T (1977) J. Org. Chem. 42: 2326
21. Rühlmann K: Synthesis 1971: 236; Saläun J, Marguerite J. (1985) Org. Synth. 63: 147
22. Saläun J (1983) Chem. Rev. 83: 619
23. Nakamura E, Sekiya K, Kuwajima I (1987) Tetrahedron Lett. 28: 337
24. Tamaru Y et al. (1987) 54th Annual Meeting of the Chemical Society of Japan, Tokyo, 1987; 3IIIM32.
25. Wailes PC, Coutts RSP, Weigold H (1974) Organometallic chemistry of titanium, zirconium, and hafnium, chap 2 Academic, New York
26. Murai S, Aya T, Renge T, Ryu I, Sonoda N (1974) J. Org. Chem. 39: 858; Ryu I, Murai S, Otani S, Sonoda N: Tetrahedron Lett. 1977: 1995
27. Oshino H. (1984) Master's thesis submitted to Tokyo Institute Technology
28. Nakamura E, Kuwajima I (1984) J. Am. Chem. Soc. 106: 3368; Nakamura E, Kuwajima I (1988) Org. Synth. 66: 43

29. Nakamura E, Aoki S, Sekiya K, Oshino H, Kuwajima I (1987) J. Am. Chem. Soc. 109: 8056
30. Harrison PG, King TJ, Healy MA (1979) J. Organomet. Chem. 182: 17
31. Ryu I, Murai S, Sonoda N (1986) J. Org. Chem. 51: 2389
32. Jorgensen JJ (1970) Organic Reactions 18: 1; Negishi E, Bagheri V, Chatterjee S, Luo F-T, Miller JA, Stoll AT (1983) Tetrahedron Lett. 24: 5181
33. Oshino H, Nakamura E, Kuwajima I (1985) J. Org. Chem. 50: 2802
34. Ryu I, Rhee I, Ryang M, Omura H, Murai S, Sonoda N (1984) Synthetic Commun. 14: 1174
35. Reetz M (1986) Organotitanium reagents in organic synthesis, Springer, Berlin Heidelberg New York
36. Boersma J (1982) In: Wilkinson G (ed) Comphrensive organometallic chemistry Pergamon, Oxford p 823
37. Giese B, Horler H, Zwick W (1982) Tetrahedron Lett. 23: 931; Giese B, Horler H (1983) Tetrahedron Lett. 24: 3221
38. Corey EJ, Boaz NW (1985) Tetrahedron Lett. 26: 6015, 6019; Alexakis A, Berlan J, Besace Y (1986) Tetrahedron Lett. 27: 1047; Horiguchi Y, Matsuzawa S, Nakamura E, Kuwajima I (1986) Tetrahedron Lett. 27: 4025; Nakamura E, Matsuzawa S, Horiguchi Y, Kuwajima I (1986) Tetrahedron Lett. 27: 4029; Johnson CR, Marren TJ (1987) Tetrahedron Lett. 28: 27
39. Horiguchi Y, Nakamura E, Kuwajima I (1986) J. Org. Chem. 51: 4323
40. Nakamura E, Kuwajima I (1986) Tetrahedron Lett. 27: 83
41. Stang PJ, Hanack M, Subramanian LR: Synthesis, 1982: 85
42. Caine D, Frobese AS: Tetrahedron Lett. 1978: 883
43. Goswami R, Corcoran DE (1982) Tetrahedron Lett. 23: 1463
44. Goswami R, Corcoran DE (1983) J. Am. Chem. Soc. 105: 7182
45. Goswami R (1985) J. Org. Chem. 50: 5907
46. Sano H, Tanaka Y, Migita T: 50th Annual Meeting of the Chemical Society of Japan, Tokyo, 1985; 2V02
47. Sano H., Suzuki T, Migita T: 50th Annual Meeting of the Chemical Society of Japan, Tokyo, 1985; 2V03
48. Nishiyama H, Matsumoto M, Matsukura T, Miura R, Itoh K (1985) Organometallics 4: 1911
49. Tamaru Y, Ochiai H, Nakamura T, Tsubaki K, Yoshida Z (1985) Tetrahedron Lett. 26: 5559; Tamaru Y, Ochiai H, Nakamura T, Yoshida Z (1986) Tetrahedron Lett. 27: 955
50. Ito Y, Fujii S, Saegusa T (1976) J. Org. Chem. 41: 2073; Ito Y, Sugaya T, Nakatsuka M, Saegusa T (1977) J. Am. Chem. Soc. 99: 8366; Ito Y, Fujii S, Nakatsuka M, Kawa-moto F, Saegusa T (1988) Org. Synth. Coll. Vol. 6: 327
51. Ryu I, Ando M, Ogawa A, Murai S, Sonoda N (1983) J. Am. Chem. Soc. 105: 7192
52. Stork G, Davies JE, Meisels A (1963) J. Am. Chem. Soc. 85: 3419
53. Ryu I, Ogawa A, Sonoda N: J. Chem. Soc., Jpn. 1985: 442
54. Ryu I, Kameyama Y, Matsumoto K, Murai S, Sonoda N (Unpublished work cited in Ref. [2])
55. Ryu I, Suzuki H, Ogawa A, Kambe N, Sonoda N (1988) Tetrahedton Lett. 29: 6137
56. Aoki S (unpublished results at Tokyo Institute of Technology)
57. Aoki S, Fujimura T, Nakamura E, Kuwajima I (1988) J. Am. Chem. Soc. 110: 3296; Aoki S, Nakamura E, Kuwajima I (1988) Tetrahedron Lett. 29: 1541
58. Rousseaux G, Sloughui N (1984) J. Am. Chem. Soc. 106: 7283
59. Rubottom GM, Marrero R, Krueger DS, Schreiner JL: Tetrahedron Lett. 1977: 4013
60. Rubottom GM, Beedle EC, Kim CW, Mott RC (1985) J. Am. Chem. Soc. 107: 4230
61. Bishop KC III (1976) Chem. Rev. 76: 461
62. Mann BE, Taylor BF (1981) ^{13}C NMR Data for Organometallic Compounds, Academic, London
63. Nakamura E, Kuwajima I (1985) J. Am. Chem. Soc. 107: 2138
64. Carmely S, Kashman Y (1981) Tetrahedron 37: 2397

Gem-Dihalocyclopropanes in Organic Synthesis

Rafael R. Kostikov[1], Alexandr P. Molchanov[1] and Henning Hopf[2]

1 Department of Organic Chemistry, University of Leningrad, University pr., 2   198904, Leningrad, Petrodvorez, USSR
2 Institut für Organische Chemie, Universität Braunschweig, Hagenring 30, D-3300 Braunschweig, FRG

Table of Contents

Gem-Dihalocyclopropanes belong to the most readily available cyclopropane derivatives known today. They have been shown to be extremely valuable starting materials for the preparation of cyclopropanes and cyclopropenes, they may be converted to bicyclobutane derivatives and spiropentanes, can lead to allenes and the higher cumulenes, cyclopentenes and cyclopentadienes, and many other classes of compounds, both hydrocarbon systems and derivatives with valuable functional groups. The article summarizes the preparative developments in the area of *gem*-dihalocyclopropane chemistry during the last decade.

Topics in Current Chemistry, Vol. 155
© Springer-Verlag Berlin Heidelberg 1990

1 Introduction

Although the first *gem*-dihalocyclopropane — 1,1-dichlorocyclopropane — was described nearly a century ago [1], this class of organic halogen compounds played no role in preparative organic chemistry for many decades. In fact, even Doering and Hoffmann's 1954 paper [2] describing the generation of dichlorocarbene from chloroform under anhydrous conditions and its subsequent trapping by olefins to provide 1,1-dichlorocyclopropanes in an experimentally simple way for the first time was of no immediate consequence for synthetic organic chemistry. The process played a role for the preparation of bi- and polycyclic hydrocarbons, i.e. the two chlorine substituents were regarded as a necessary evil to get rid off as quickly as possible, rather than as functional groups in their own right to be used for further modifications in subsequent steps. This application of the *gem*-dihalocyclopropane group is still very popular [3], as will be shown by some examples discussed later. However, during the last two decades the great preparative potential of cyclopropanes of all types has become more and more apparant. In fact, these three-membered ring systems have been described as "pseudofunctional groups" [4].

The published experimental material has thus accumulated rapidly, and a number of useful and comprehensive reviews have already appeared [5–10]. In particular, Weyerstahl has published an extensive review of articles published between 1967 and 1980 dealing with the preparation and the chemical behavior of dihalocyclopropanes [11]. The need for a further summary of this part of cyclopropane chemistry may thus not appear obvious. However, besides presenting work published since then, the present article strictly concentrates on the reactive behavior of the title compounds. Neither their preparation nor their spectroscopic, especially their NMR properties will be discussed here, since these questions are dealt with in depth in Weyerstahl's review [11]. For a summary of the work published prior to 1970, Wendisch's monumental contribution to the Houben-Weyl-Müller series remains unchallenged [10]. As far as cycloadditions of dichlorocarbene to olefins are concerned, the monograph by Zefirov, Kazimirchik and Lukin [12] is valuable (literature coverage up to 1984), although this book has not been translated into English. One of the most useful recent compilations listing the additions of all types of halocarbenes under phase transfer conditions to a very large collection of unsaturated systems — including numerous polyfunctional molecules — has been provided by Keller [13]. Although this monograph lists hundreds of dihalocyclopropanes, novel derivatives and routes to prepare them appear continuously [14–16].

When two *geminal* halogen substituents are introduced into a cyclopropane ring, a distinct and consistent change of geometry is observed. In 1,1-difluorocyclopropane the bond opposite the halogen bearing carbon, i.e. the C^2-C^3-bond is lengthened (155.3 pm) relative to cyclopropane itself (151.4 pm [17, 18]), whereas the adjacent bonds are shortened. This effect corresponds to a weakening of the former (by about 8 to 10 kcal/mol) and a strengthening of the latter bond type. More complex cyclopropane derivatives like 4,4,8,8-tetrafluoro-tricyclo-$[5.1.0.0^{3,5}]$octane show analogous effects [19]. Although the experimental results for 1,1-dichlorocyclopropane are contradictory [20–22], the above tendency is preserved. Extensive theoretical work on the influence of donor and acceptor substituents on the electronic and geometric properties of cyclopropanes has been published [23, 24]; it shows the same trends for 1,1-difluoro- and 1,1-dichlorocyclopropane, i.e. a bond length increase for the opposing and a decrease for the adjacent bonds. As expected, these changes in molecular geometry, and hence bond energies, exert a strong influence on the chemical behavior of the *gem*-dihalocyclopropanes: In those reactions in which ring-opening takes place, it is the $C^2—C^3$-bond that is broken preferentially, and in most cases this process occurs with a higher reaction rate than that observed for the corresponding unsubstituted cyclopropanes. It should be pointed out, though, that in many reactions of the *gem*-dihalocyclopropanes the integrity of the three-membered ring is preserved. Rather then subsuming the published experimental material under the two headings of reaction products with retained or destroyed cyclopropane moiety we prefer to discuss the reactions of the *gem*-dihalocyclopropanes according to reaction types.

2 The Reduction of *Gem*-Dihalocyclopropanes

The catalytic hydrogenation of 7,7-dihalonorcaranes proceeds with opening of the three-membered ring and leads ro methylcyclohexane, whereas the reduction with sodium in alcohols affords norcarane [2]. Hydrogenolysis of 1,1-difluoro-2-methyl-3-phenylcyclopropane in the presence of a Ni—Re or Pd catalyst at atmospheric pressure provides a mixture of 1-phenyl-butane, 1-phenyl-2-fluorobutane, and 1-phenyl-2,2-difluorobutane, respectively [25].

When the *gem*-dihalocyclopropanes are reduced by dissolving alkali metals in alcoholic solvents or in liquid ammonia the strained ring is retained.

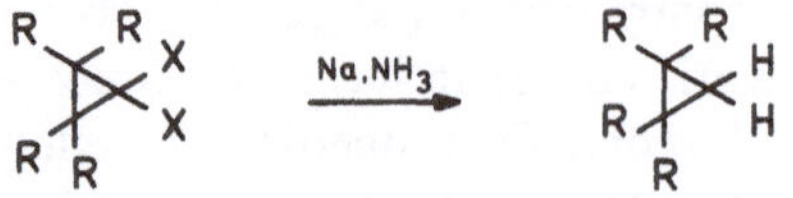

Rafael R. Kostikov et al.

ref. 26,27

R = H, Alk, Ar
X = Cl, Br

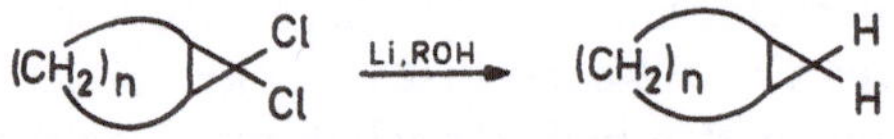

ref. 28

The former reduction is thought to occur by two single electron transfers (SET) from the metal surface to the cyclopropane derivative providing a halide and cyclopropyl anion initially. The latter is protonated by the solvent thus leading to the monohalogen derivative which can undergo the reduction process for a second time.

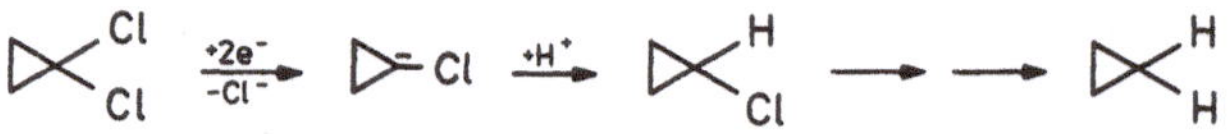

With sodium in liquid ammonia only the fully reduced compounds are obtained, even if the conversion is kept low. To explain the absence of the monohalide, it has been proposed that a cyclopropylidene radical anion is formed under these conditions, which is transformed to the dehalogenated product directly [29, 30].

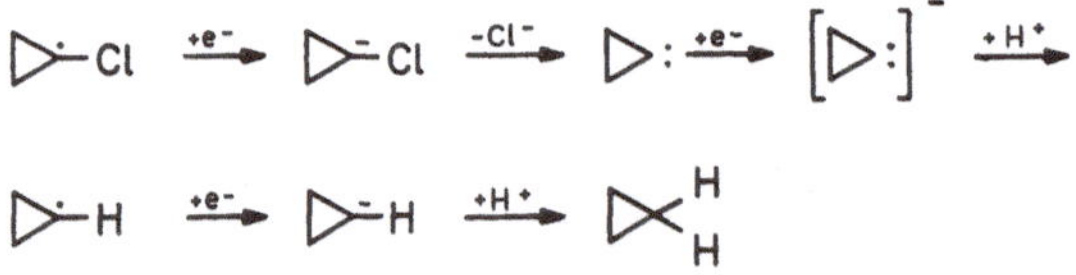

The reduction of 1-alkyl-8,8-dibromo-bicyclo[5.1.0]octanes by methyl magnesiumbromide in tetrahydrofuran provides mixtures of fully and partially reduced products [31]. A comparable result is obtained when zinc in dimethylformamide is employed; as a side reaction the formation of allenes is observed in this case (see below) [32].

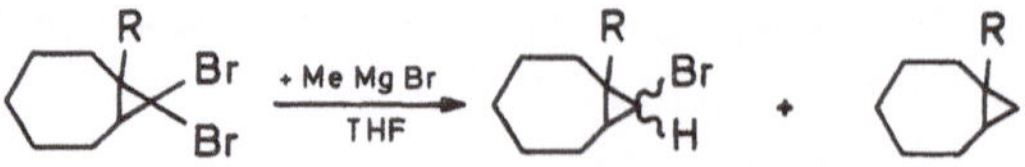

ref. 31

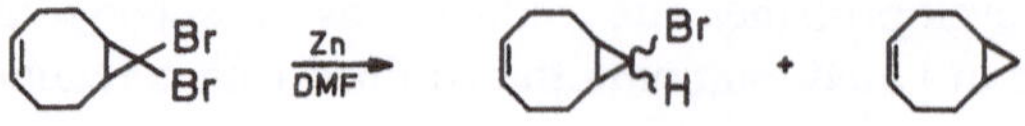

ref 32

44

To increase the selectivity of the process and at the same time improve the yields of the partially reduced products many reduction systems have been studied: Zn/ROH [33], Mg/ROH—THF [34], Al/ROH—NaOH [35], LiAlH$_4$ [36], (n-C$_4$H$_9$)$_3$SnH [37, 38], sodium bis(2-methoxyethoxy)aluminium hydride (SMEAH) [39], to name but a few. A simple and effective way to convert dibromocyclopropanes into their monosubstituted derivatives consists in the use of non-metallic reagents: diethyl phosphite in the presence of triethylamine [40], sodium dithionite in aqueous dimethylformamide [41], and a number of other reducing reagents like Na$_2$S$_2$O$_3$, Na$_2$SO$_3$, Na$_2$S, Na$_2$HBO$_2$ [40]. Under these conditions the monobromides are obtained in up to 80% yield, with less than 10% of the fully reduced cyclopropanes. The stereoselectivity of the process is not very high, and depends both on the structure of the starting dibromide and the reduction conditions employed. Thus the ratio of the isomeric monobromides in the above scheme is 1:1 if zinc in ethanol is used [33], 1.7:1 (*cis: trans*) with aluminum in alcoholic alkali hydroxide [35], and only 0.11–0.35 when lithium aluminumhydride is employed. On the other hand, a *syn*-isomer is the main product with tri-*n*-butyltinhydride [38].

The reduction of the dibromo derivatives is thought to occur via zinc organic species [42], whereas radical intermediates are apparently involved in the reduction of the monohalogen compounds [43].

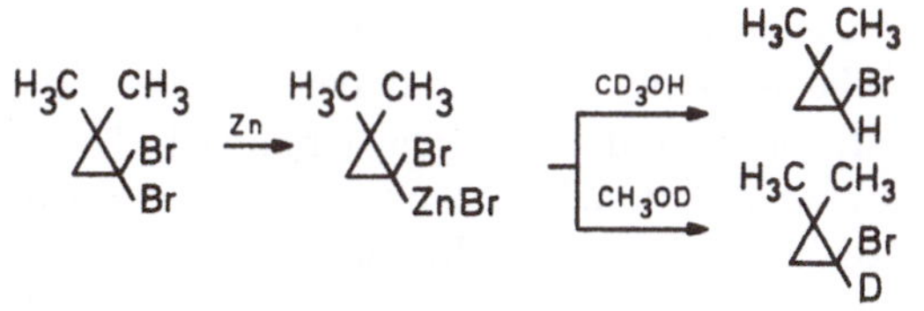

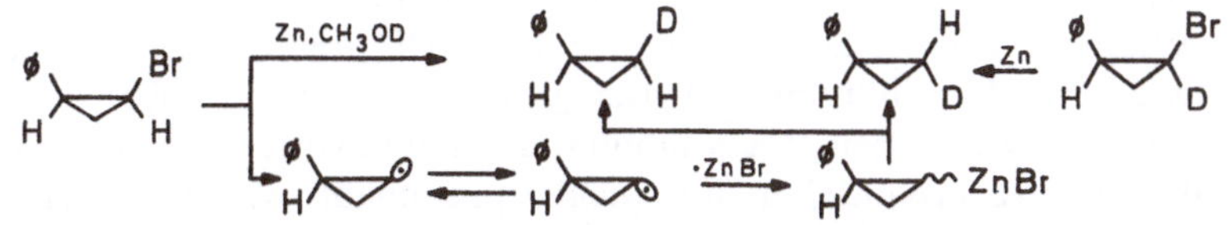

The selectivity of the reduction of dichlorocyclopropanes to the monohalogenated compounds is normally quite low; if, however, LiAlH$_4$/(C$_6$H$_5$)$_3$SnH is employed the monochloro derivatives are obtained in good yields [44–47]. A stereoselective monoreduction of *gem*-dichlorocyclopropanes has recently been accomplished with potassium diphenylphosphide. For example, when 7,7-dichlorobicyclo[4.1.0]heptane is stirred with this reagent in dimethylsulfoxide *exo*-7-chlorobicyclo[4.1.0]heptane is formed in 85% yield, and only small amounts of its *endo*-isomer are produced as side-product [48 a]. *gem*-Dibromocyclopropanes have been reduced stereoselectively by empoying potassium dimethylphosphite [KP(O) (OMe)$_2$] [48 b]. The reduction of fluorochlorocyclopropanes with tri-*n*-butyltinhydride proceeds chemoselectively leading to the corresponding fluorocyclopropanes [49, 50]. The stereoselectivity of the process appears to depend on the ability of the substituent to stabilize a radical center (CH$_3$ > H > OCH$_3$).

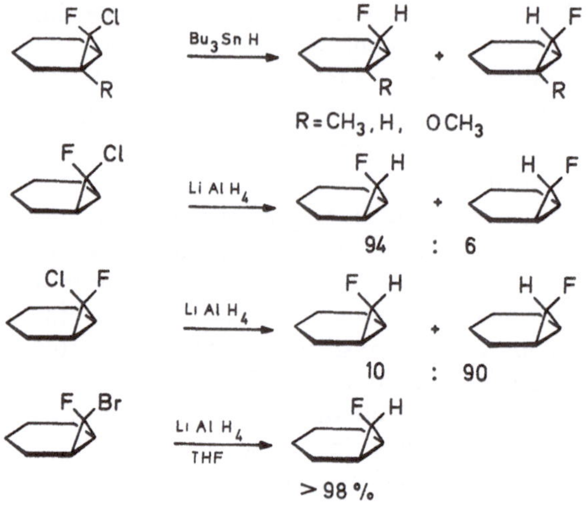

R = CH₃, H, OCH₃

The reduction of bromofluorocyclopropanes with LiAlH₄ in tetrahydrofuran at 65 °C proceeds stereoselectively with retention of configuration and formation of the monofluorocyclopropanes [50]. A similar result has been obtained for chlorofluorocyclopropanes (diglyme, 100 °C) [44]. A configurationally stable carbanion has been invoked as the reaction intermediate in the latter case. When the dihalocyclopropane substrate contains another reducible reaction center, the chemoselectivity of the reduction process depends — as might have been expected — on the nature of the reducing agent employed. Thus, in dibromocyclopropanes containing an additional keto or imino function it is the latter group which is reduced by lithium aluminumhydride [51, 52]. On the other hand, the reduction by tri-*n*-butyltinhydride or zinc in ethanol yields the monobromocyclopropanes [53]. A complex reaction mixture is formed when dibromocyclopropanoic esters are subjected to LiAlH₄-reduction, since both the ester function and the dibromomethylene group of the three-membered ring participate in the process. When excess lithium aluminumhydride is employed even allenic alcohols are produced in low yields, presumably via cyclopropylidene intermediates (cf. Sect. 3) [54].

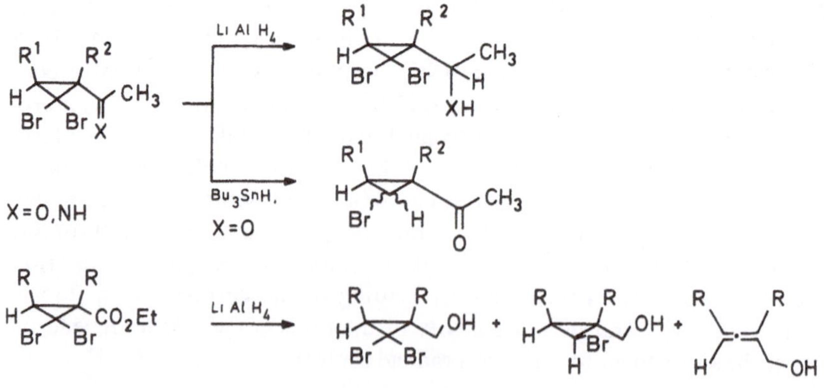

3 Reactions with Organolithium Compounds

Lithium metal or alkyllithium derivatives react with dihalocyclopropanes to provide the corresponding lithiohalocyclopropanes I which are stable at temperatures around −100 °C. These metalated species are easily trapped with electrophiles (R—X) like methyl or ethyl iodide, trimethylstannyl chloride, trimethylsilyl chloride etc. In the case of the unsaturated bicyclic substrate II a double bond migration is observed, which in the presence of excess starting bromide is accompanied by isomerization of the *exo*-lithio intermediate III to its *endo*-isomer IV [58].

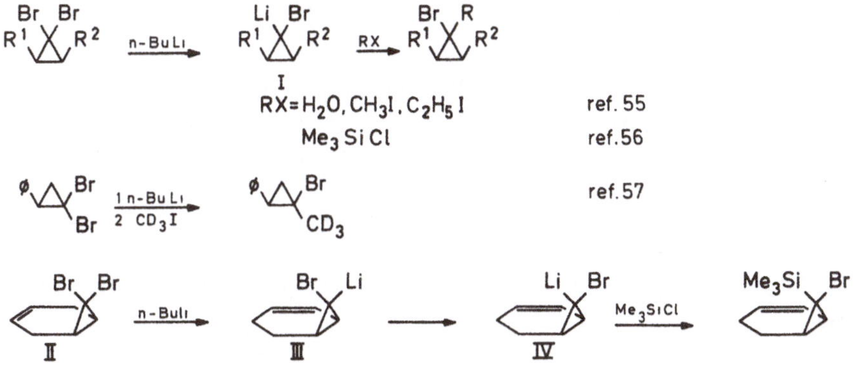

If 7,7-dibromonorcarane is metalated with *n*-butyllithium and the resulting intermediate trapped with CH$_3$OD the *endo*-lithio species Va furnishes only the *exo*-7-bromo-7-deuterionorcarane VI, whereas the epimeric organolithium compound Vb provides a mixture of both diastereomeric monomromides and an alkylation product [59].

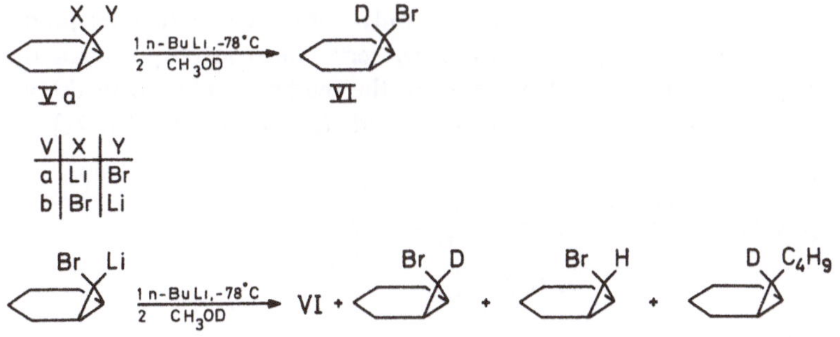

V	X	Y
a	Li	Br
b	Br	Li

The trapping of lithiohalocyclopropanes by carbon dioxide, aldehydes and acid chlorides, respectively, constitutes a useful route to the corresponding cyclopropanecarboxylic acids, alcohols and ketones. In the case of ketones an intramolecular loss of lithium bromide may take place yielding spiroepoxides which in turn may be isomerized to cyclobutanones.

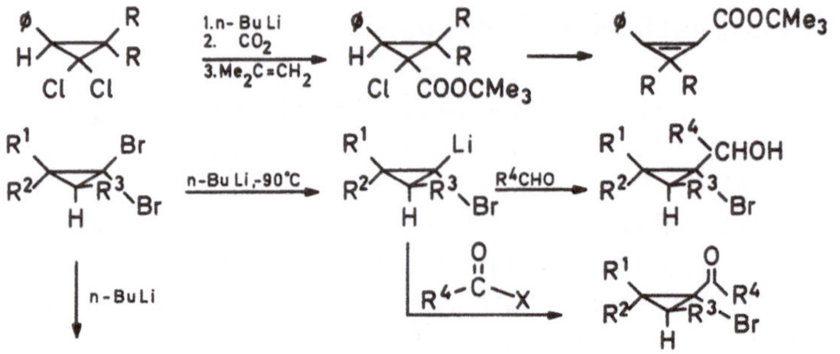

ref. 60,61

ref. 62,63

ref. 64

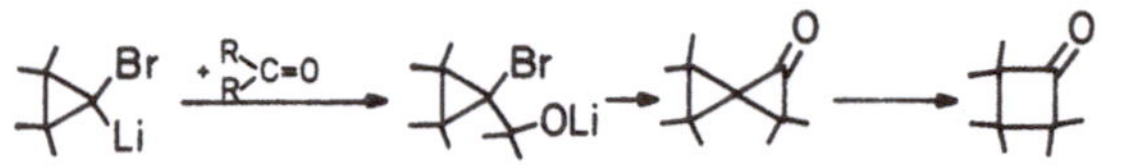

ref. 65

At temperatures above −100 °C the lithiohalocyclopropanes are converted to cyclopropylidenes by formal loss of one molecule of lithium halide. The main route of stabilization of these "carbenacyclopropanes" consists in the formation of allenes. The process which has been termed the Doering-Moore-Skattebøl allene synthesis ("DMS-synthesis") [66–68], has been developed into the most general method for the preparation of these reactive compounds which especially during the last decade have been used in organic synthesis with growing success [69, 70].

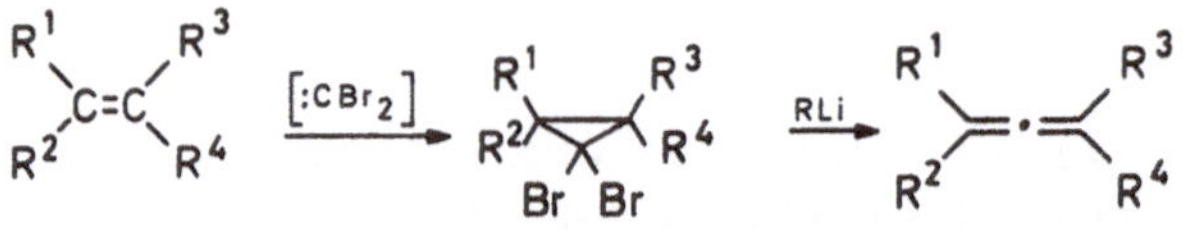

Formally, in the two steps of the DMS-process (dibromocyclopropanation and reaction with alkyllithium) a carbon atom is inserted between the two centers of a double bond. The reaction may be extended to the preparation of still higher cumulated bond systems as well as to numerous other — including functionalized — allenic systems which cannot or only with much effort be prepared by other routes. The examples shown here serve illustrative purposes only, for more extensive coverage of the literature the reader is referred to the various reviews and monographs which have appeared recently [66, 69, 71, 72].

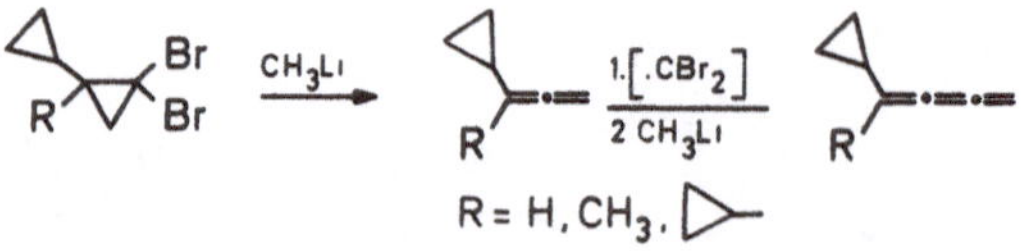

ref. 73–76

ref. 77

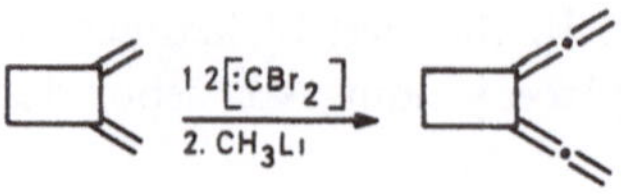

ref. 78

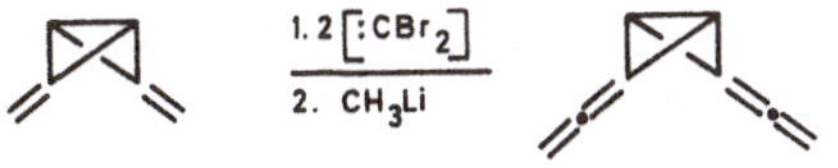

ref. 79

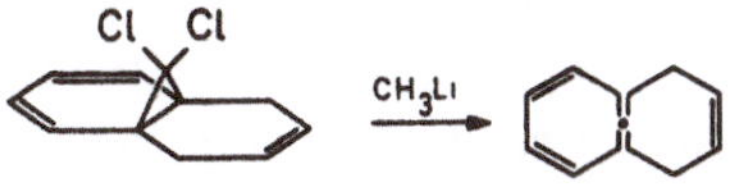

ref. 80

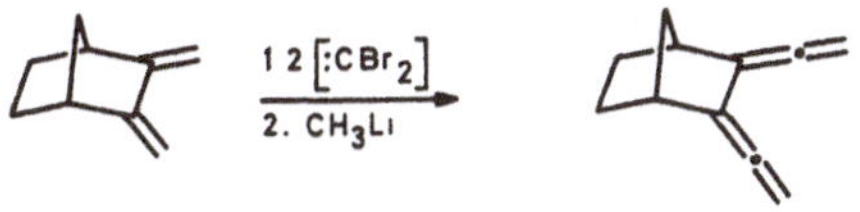

ref. 81

Occasionally the cyclopropylidene to allene isomerization cannot take place for structural reasons. If, for example, the expected allene would be very highly strained, as is the case for certain cyclic allenes, then the reaction is forced to follow an alternative path. A case in point is provided by 1-alkyl-7,7-dibromonorcaranes which undergo a carbene insertion reaction when treated with methyl lithium to provide bicyclobutanes rather than allene derivatives.

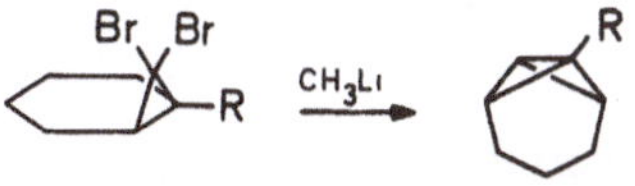

R=H   ref. 82
R=CH(CH$_3$)$_2$
C(CH$_3$)$_3$ ref. 83

A more complex variant of this reaction has been used by Paquette in his short (six-step) synthesis of heptalene from naphthalene [84].

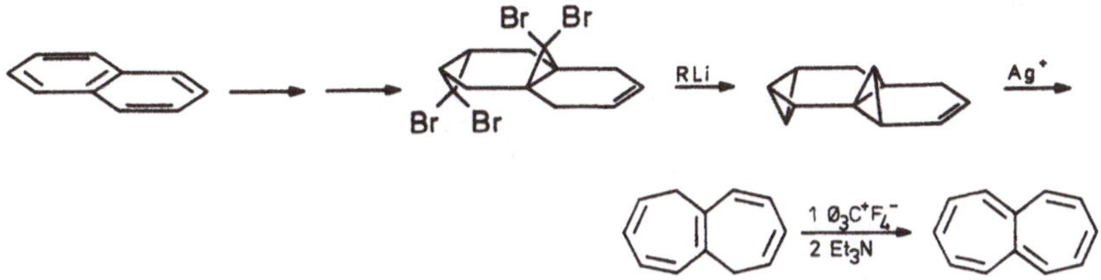

In another series of experiments involving tricyclic dibromocyclopropanes Warner and co-workers have studied the behavior of a propellane derivative towards methyllithium either in the presence or absence of trapping agents. Whereas in the former case a dimer is produced, with diphenylisobenzofuran (DPIBF) two adducts are obtained in a 2:1 ratio in 24% yield [85].

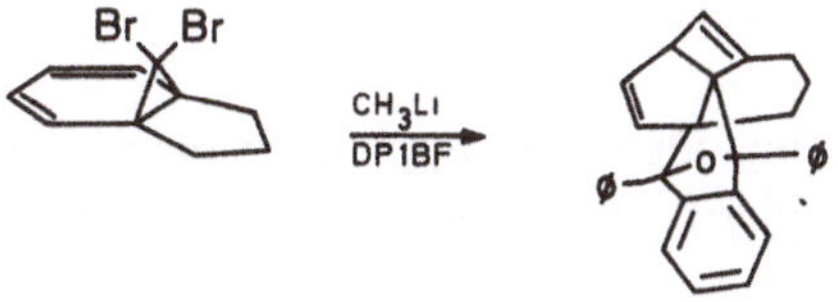

Rafael R. Kostikov et al.

According to the authors an allene intermediate is not involved in this complex transformation [85]. If true, allenes are also unlikely to be produced in the dehalogenation reactions of some related dihalocyclopropanes [86].

That the allene route should always be kept in mind, though, is demonstrated for example by 6,6-dibromobicyclo[3.1.0]octane and 8,8-dibromo-bicyclo[5.1.0]-octane, respectively. When these bicyclic dibromides are reacted with methyl-lithium at room temperature they are evidently converted into the corresponding cyclic allenes since these intermediates may be either trapped by reagents like styrene or dimerize, [2+2]adducts being formed in both cases.

The successful application of the Doering-Moore-Skattebøl reaction to the preparation of relatively small-ring heteroorganic cycloallenes is of recent origin. A specific example is provided by 6,6-dichloro- and 6,6-dibromo-3-oxa-bicyclo-[3.1.0]hexane, respectively, which yield 1-oxa-3,4-cyclohexadiene when treated with n-butyllithium at low temperatures. The formation of the allene was proven by various trapping experiments [89].

The trick of stabilizing reactive molecules by bulky substituents has often been exploited in organic chemistry. It has also been employed to make certain strained cycloallenes isolable compounds. In fact, the 1,2-cyclooctadiene shown is the first kinetically stable cycloallene of this type [88].

The allene generating reaction has also been applied to more complex bi- and polycyclic substrates. Two examples from the recent literature suffice to illustrate this point.

50

ref.90

ref.91

No other dibromocarbene adduct has probably been used in more subsequent reactions than that of benzvalene. Besides the allene producing reaction shown, Christl and his students have described dozens of transformations of this versatile *gem*-dibromocyclopropane, among them practically all reaction-types discussed in the different chapters of this review [92].

Furthermore, this approach has been applied to the synthesis of substituted bicyclobutanes. For example, tetra-substituted dibromocyclopropanes carrying one or more aromatic substituents provide the diphenyl-bicyclo[1.1.0]butanes shown when treated with methyllithium.

ref.93

The cyclopropylidene intramolecular carbene insertion is not restricted to carbon hydrogen bonds; how carbon-halogen bonds may participate is illustrated by the following examples. Treatment of 1-chloromethyl-1-methyl-2,2-dibromo-cyclopropane with methyllithium leads preferentially to 1-methylbicyclobutane [94], and a particularly interesting and important carbon-halogen insertion takes place during the synthesis of [1.1.1]propellane from 2-chloromethyl-3-chloropropene [95–98]. Once the distance between the carbene center and the halogen atom becomes too large allene formation takes over again [94].

Substituted cyclopropylidenes have been shown to participate in both inter- and intramolecular addition reactions with olefins. The resulting products are spiropentane derivatives as well as carbene dimers which are formed as side-products [99, 100]. In the absence of olefinic reaction products the latter may even become the main products [99 b].

51

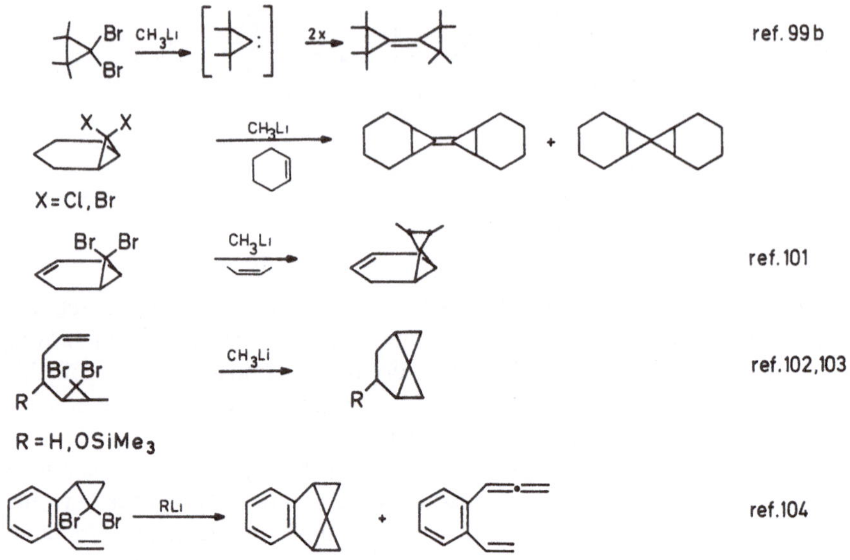

ref. 99 b

X = Cl, Br

ref. 101

ref. 102,103

R = H, OSiMe$_3$

ref. 104

When the bis-dibromocarbene adduct of *o*-divinylbenzene is treated with methyl lithium a surprising influence of the reaction temperature on the product composition is noted.

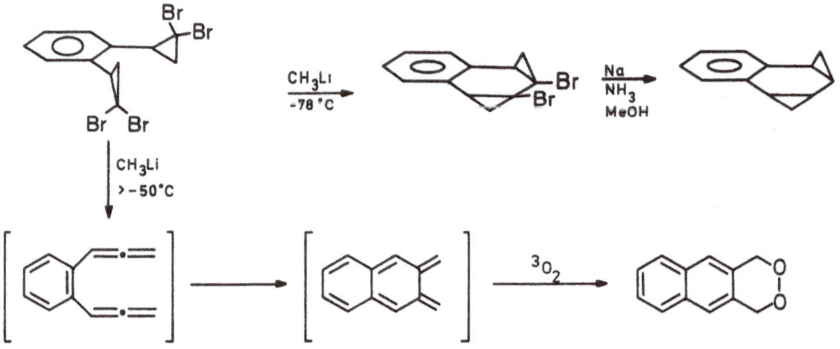

Whereas below −50 °C a ring closure reaction by 1,6-bromine elimination takes place to give — after reduction — the *trans*-bis-homobenzene derivative shown, at higher temperatures the typical allene forming process is observed. The resulting aromatic bisallene undergoes further cyclization providing an *ortho*-quinodimethane which may be trapped to an *endo*-peroxide by addition of oxygen [105].

If the double bond is connected directly to the three-membered ring, i.e. if the substrate is a dihalo-vinylcyclopropane derivative, a vinylcyclopropylidene to cyclopentadiene isomerization, the so-called Skattebøl rearrangement, takes place upon treatment with alkyllithium. Again, this ring-forming step competes with a ring-destruction process leading to allenic hydrocarbons (vinylallenes, π-systems which possess both cumulated and conjugated diene subsystems and which are preparatively quite useful [106, 107]).

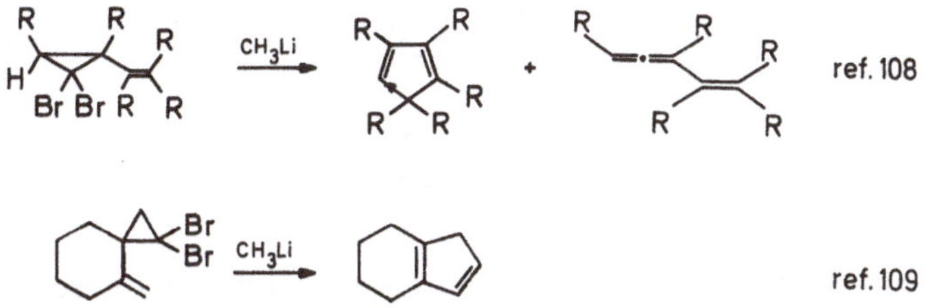

ref. 108

ref. 109

Mechanistically this process is interesting since it involves a carbene-carbene rearrangement as was shown by Skattebøl [110], Baird [111] and Brinker [112] using isotopically labeled substrates.

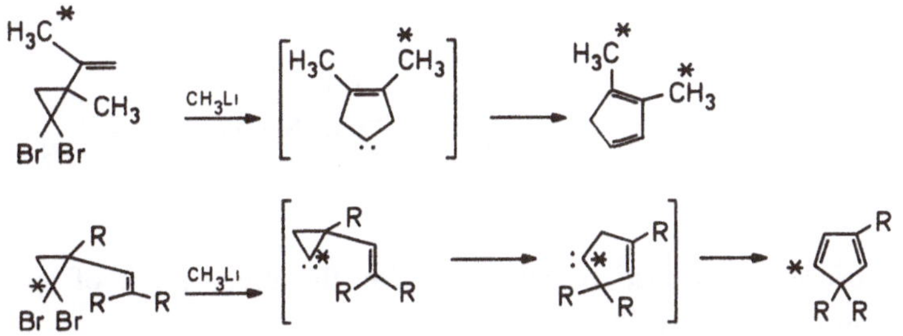

A vinylcyclopropylidene to cyclopentenylidene isomerization evidently also taken place when 7,7-dibromo-bicyclo[4.1.0]hept-2-ene and 8,8-dibromo-bicyclo[5.1.0]-oct-2-ene are treated with methyllithium. The former compound is converted to 7-bromo-7-methyl-norbornene by intermolecular insertion into the C—Br bond of methyl bromide, whereas tetrahydropentalene is obtained from the latter [113].

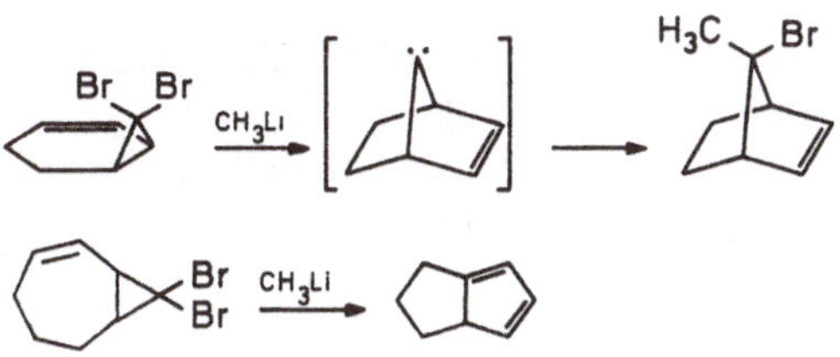

Skattebøl rearrangements in more complex dibromo-vinylcyclopropanes have also been reported. Thus, either a system consisting formally of two double bonds and one dibromocyclopropane unit or an educt containing two dibromo-cyclopropane moieties and one double bond undergo several carbene-carbene isomerizations upon treatment with methyllithium. In both cases, complex product mixtures arise.

53

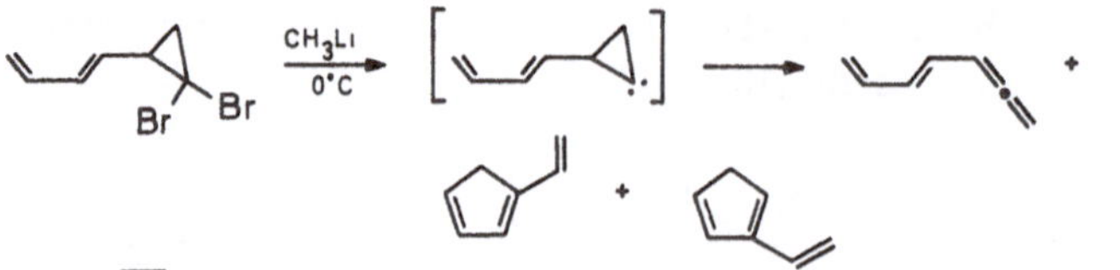

ref. 114,115

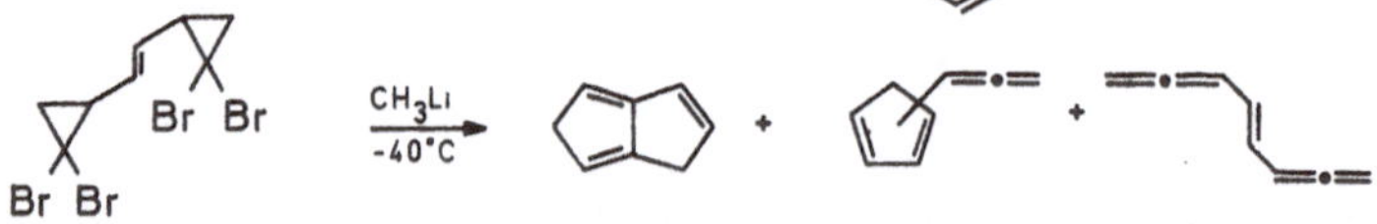

ref. 116,117

No systematic efforts have apparently been undertaken to replace the carbon atoms of the above molecules by isolated hetero atoms or hetero atom-containing fragments. That this may, in fact, be a worthwhile undertaking is demonstrated by the *tert*-butylimino derivative VII which is converted to a pyrrole (plus the other products shown in the scheme) by methyllithium.

Presumably, the vinylcyclopropylidene to cyclopentadiene isomerization is involved again.

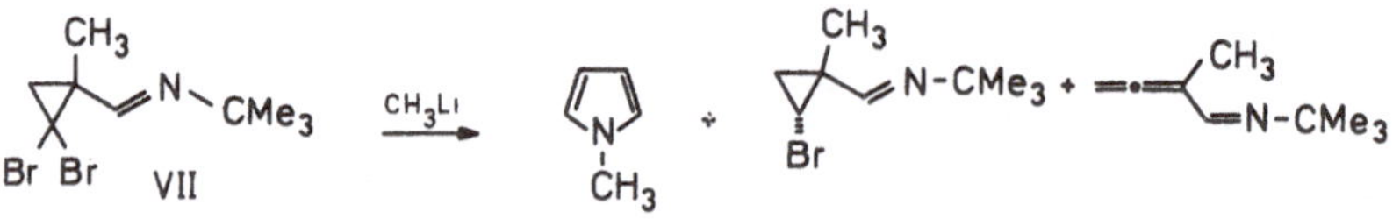

ref. 118

This process may also be employed for the preparation of specifically ring-substituted pyrroles [119].

Novel preparative possibilities arise when cyclopropylidene intermediates are generated from functionalized dihalocyclopropanes. Thus Baird has shown that bicyclic ethers may be formed in high yield by an intramolecular insertion from cyclopropyl alkyl ethers [120].

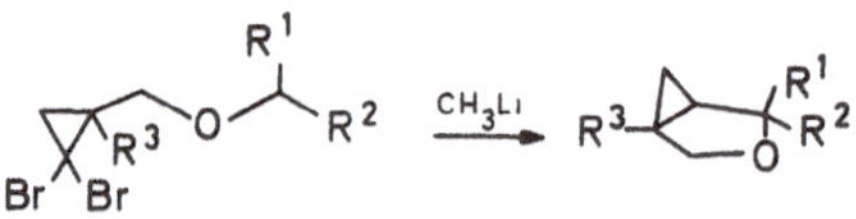

Similar processes have been reported for other ethers [121], alcohols [122], amines [123], sulfides [124], and acetals [125].

Concerning the latter class of compounds a carbene insertion has been employed to prepare the major constituent of the sex attractant of the olive fly (*Dacus oleae*).

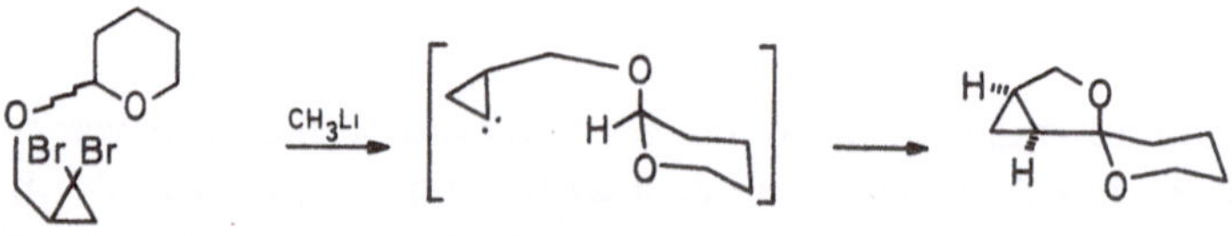

Interestingly, no allenic products are formed during this reaction. The two oxygen atoms seem to prevent a competing cyclopropylidene to allene isomerization, at least at the reported reaction temperature of −75 °C [126].

The preparative versatility of the dibromovinylcyclopropanes is further demonstrated by their oxidation to *gem*-dibromo-cyclopropylaldehydes, compounds which are difficult to obtain by other routes [127].

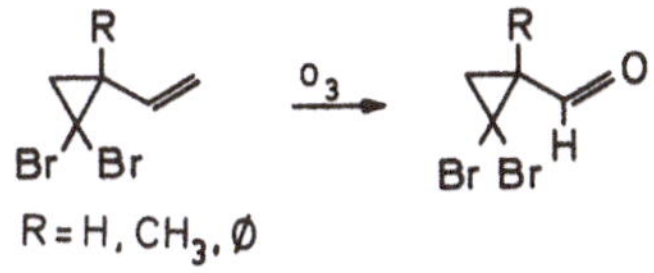

R = H, CH$_3$, Ø

Over-all, this route allows the conversion of a conjugated diene into a highly reactive building block which may be subjected to further transformations [128]. The direct oxidation of the vinylcyclopropanes to the corresponding *gem*-dibromocyclopropanoic acids has also been reported [127, 128].

4 Dehydrohalogenation Reactions

When subjected to strong bases, *gem*-dihalocyclopropanes undergo dehydrohalogenations, and cyclopropenes are formed. These are generally unstable under the reaction conditions and participate in further transformations. The most common of these processes is the isomerization of the newly formed double bond from the *endo*- to the *exo*-orientation, followed by a second dehydrohalogenation step. The methylenecyclopropenes thus generated are still not stable, and subsequently tend to rearrange to less strained systems.

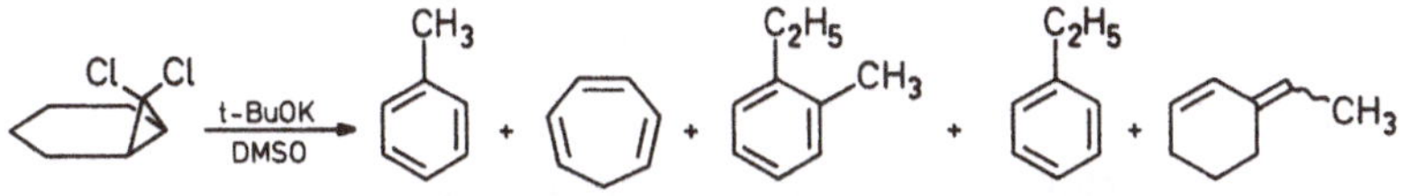

ref.129

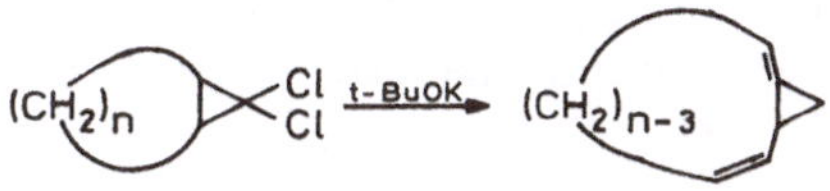

ref.130,131

ref. 130

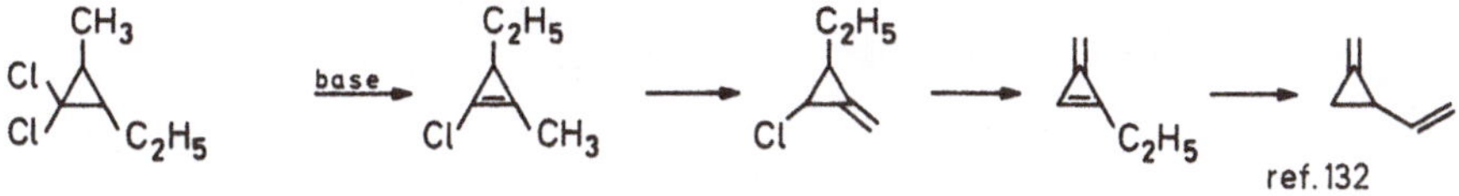

ref. 132

ref. 132

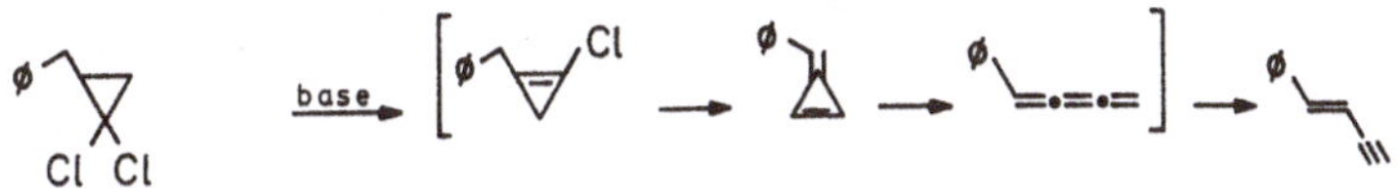

ref.133

55

Whereas 7,7-dihalonorcaranes are converted to toluene at 500 °C, they provide cycloheptatriene in yields up to 66% when heated in quinoline or dimethylaniline [134, 135].

X=Cl,Br

Tropones are obtained in good over-all yields when the dihalocarbene adducts of Δ^2- and Δ^3-norcarenes are first allylically oxidized and the resulting ketones subsequently subjected to a dehydrohalogenation sequence [136].

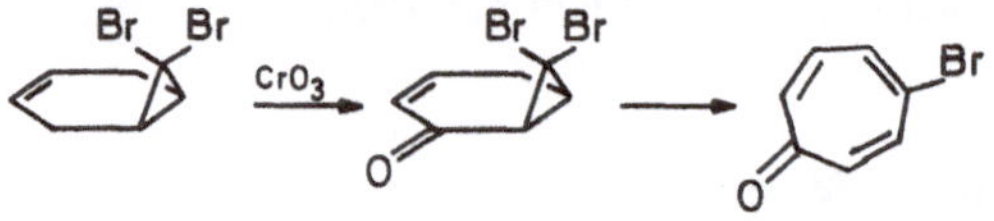

In yet another tropone synthesis Skattebøl has shown that dibromocyclopropanes carrying a phenolether function react to 11-oxa-tricyclo[5.5.0.0^{7,9}]undecatrienes when treated with methyllithium. The strained polycyclic intermediate is converted into a 2-alkenyltropone derivative when heated to about 200 °C [137].

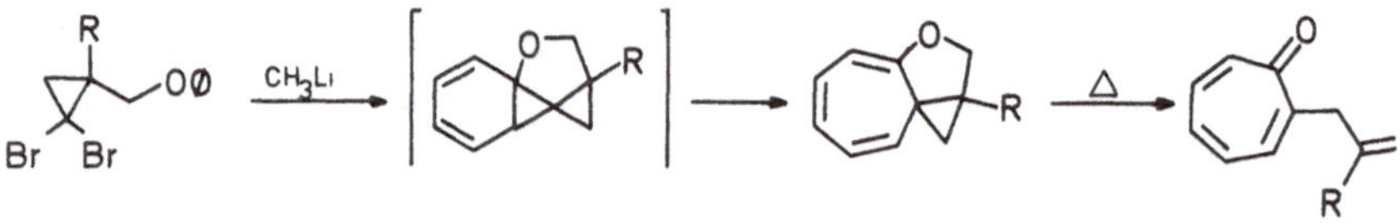

Stabilization of the cyclopropene formed during the dehydrohalogenation may also be achieved by letting its double bond become part of an aromatic system. By this approach several highly strained hydrocarbons like cyclopropabenzene, cyclopropanaphthalene etc. have been synthesized.

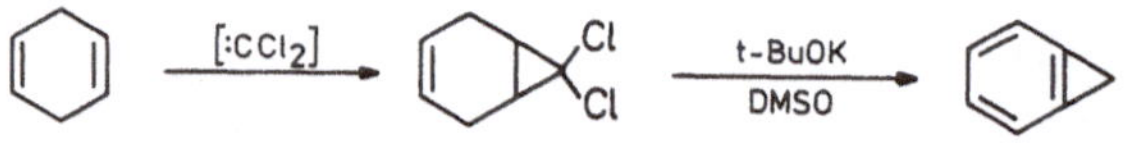

ref. 138

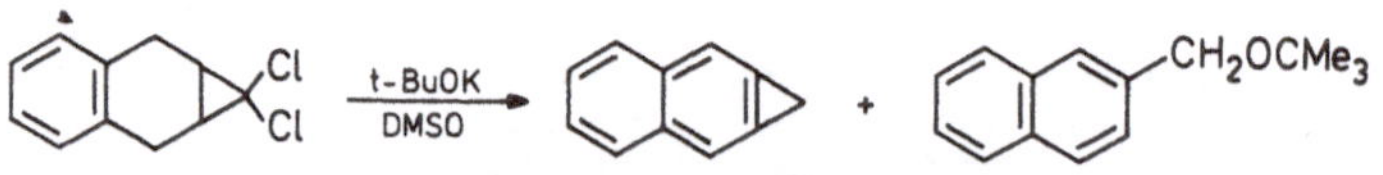

ref. 139

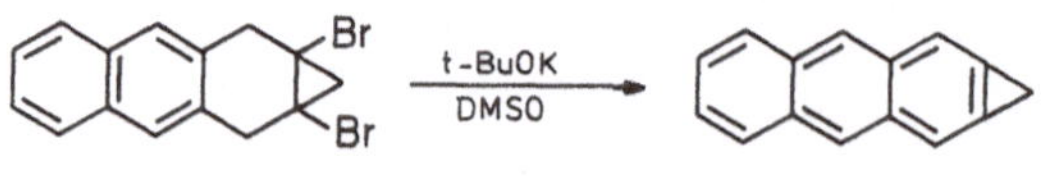

ref. 140

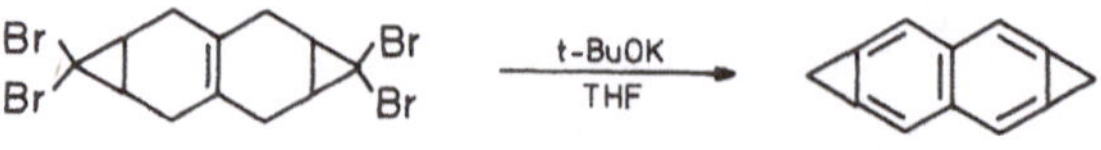

ref. 141

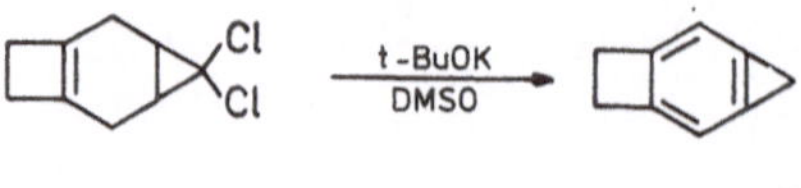

ref. 142

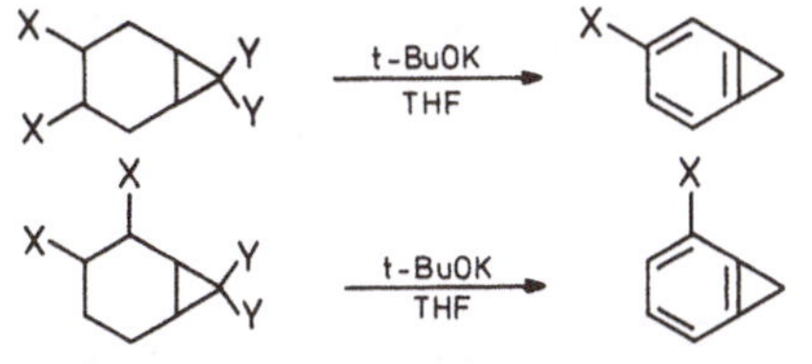

X = Cl, Br
Y = Cl, Br   ref.143,144

When cyclopropanes carrying three or more halogen substituents are treated with strong bases the initially formed halocyclopropenes may be isolated [147]. Quite generally, though, 1,1,2-trihalocyclopropanes form propargyl halides or enynes when treated with strong bases [147]. Vinylcarbenes may be intermediates in these transformations, as 1-halo-3,3-dimethylcyclopropenes and 1,2-dihalo-3,3-dimethylcyclopropenes easily undergo ring-opening to vinylcarbenes, which can be intercepted by olefins thus allowing the preparation of vinylcyclopropane derivatives [147, 148].

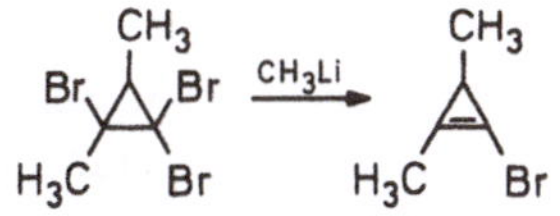

ref.145

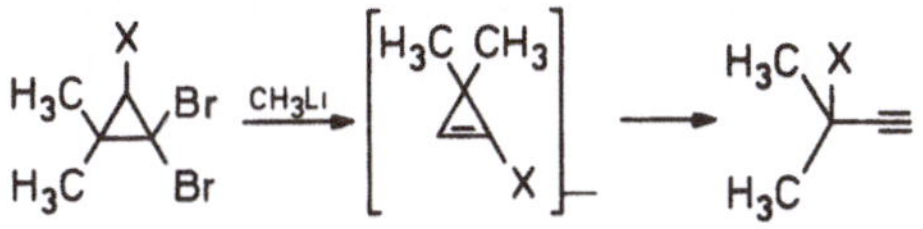

ref.145

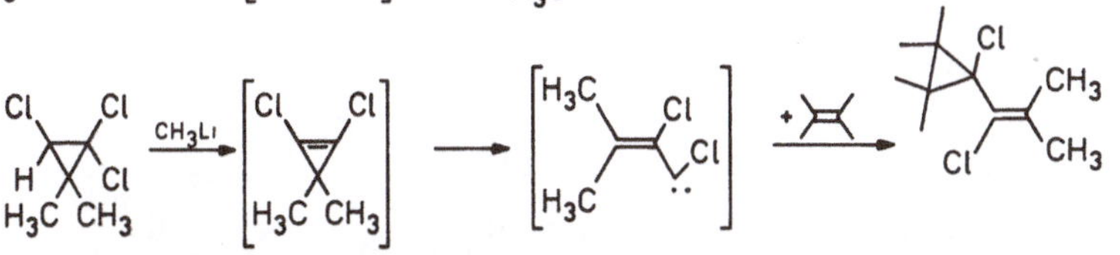

ref.146

This method has successfully been applied for the addition of chloro(1-chloro-2,2-dimethylvinyl)carbene — formed from 1,1,2-trichloro-3,3-dimethylcyclopropane by dehydrochlorination and subsequent ring-opening — onto ketene methyl silyl acetals, the products of which are transformed to interesting allenic esters when treated with tetrabutylammonium fluoride [149].

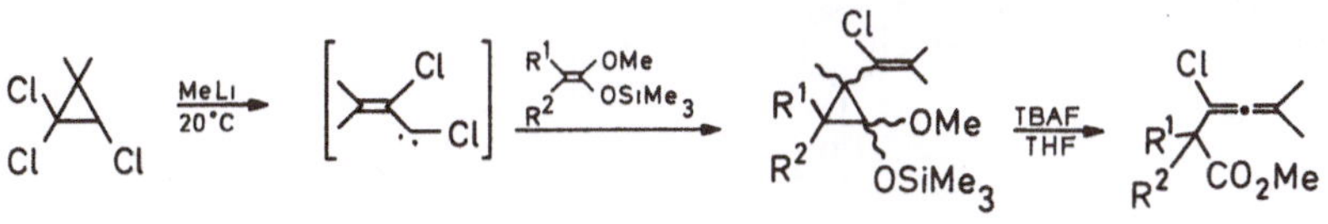

1,1,2,2-Tetrachloro-3,3-dimethylcyclopropanes with an additional substituent on one of the methyl groups have also been treated with methyllithium to yield the corresponding 1,2-dichlorocyclopropenes which ring-open to the corresponding vinylcarbenes at 0 °C. The latter add intermolecularly to olefinic substrates giving highly substituted vinylcyclopropanes in high yields. These products in turn are quite useful for further transformations to building blocks for organic synthesis [148].

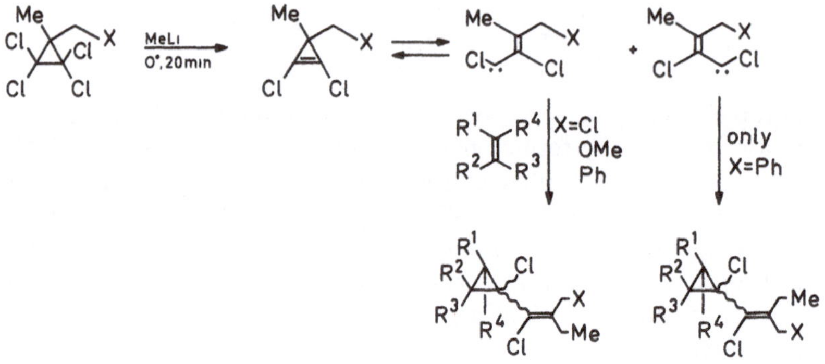

Dehydrochlorination of pentachlorocyclopropane, formed from trichloroethylene and sodium trichloroacetate as a source of dichlorocarbene, yields tetrachlorocyclopropene [150], a particularly versatile reagent for various applications. It is a reasonably reactive dienophile [151], a reagent applicable to heterocyclic synthesis [152], and an electrophile for aromatic substitutions [153] and additions to alkenes [154] in the presence of Lewis acids.

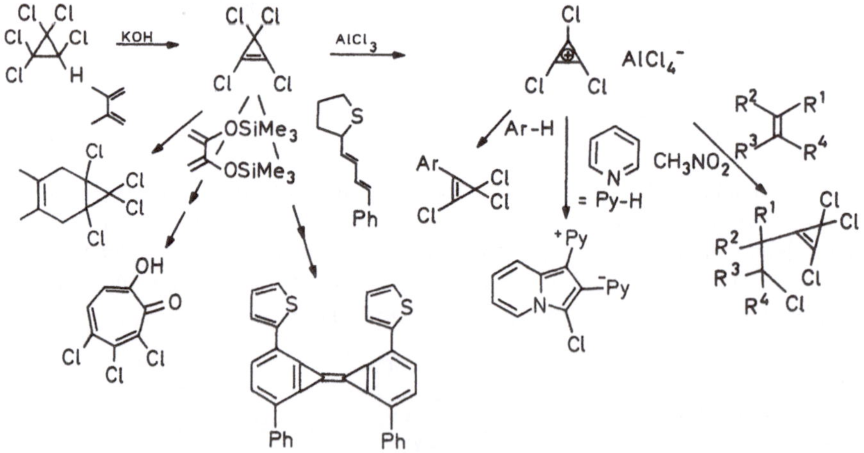

A. de Meijere and co-workers have extensively studied yet another application of tetrachlorocyclopropene [148]. When heated in the presence of olefins to temperatures between 150 and 180 °C, it ring-opens to tetrachlorovinylcarbene which very efficiently adds to the olefinic double bond providing high yields (60–90%) of the corresponding 1-chloro-1-(trichlorovinyl)cyclopropanes [148].

$$R^1R^4 \quad + \quad \text{Cl}_2\text{C}=\text{CCl}_2 \text{ (cyclopropane)}$$

150–180 °C

Li, Bu^{t}OH / THF

(*i*) 2 BuLi (*ii*) EX

KOH CH$_3$OH

(*i*) 3 BuLi (*ii*) 2 EX

E = Si(CH$_3$)$_3$

(*i*) BuLi (*ii*) E'X

H$^+$

CO$_2$CH$_3$ ← C(O(CH$_3$))$_3$

Si(CH$_3$)$_3$

Quite a wide variety of alkenes have been subjected to this carbene addition [148]; the products are multifunctional small ring molecules which may not only be reduced to simple vinylcyclopropanes, but to various substituted cyclopropyl-acetylenes and cyclopropylideneacetates which are particularly useful and versatile building blocks for organic synthesis [155].

Phase transfer-catalyzed reactions have recently been employed to dehydro-halogenate *gem*-dihalocyclopropanes [156, 157]. Thus, 1-methylene-2-vinylcyclopropane has been prepared from 1,1-dichloro-2-ethyl-3-methylcyclopropane in 60 % yield. Under the reaction conditions (solid KOH, DMSO in the presence of dibenzo-18-crown-6, 100–130 °C) further transformations may take place, however. For example, monoalkylated cyclopropanes have been converted to mixtures of acyclic enynes and conjugated trienes. And 7,7-dichloronorcarane is converted to toluene under these conditions.

Cl Cl

H$_3$C   CH$_2$CH$_3$

$\xrightarrow[\text{DB18C6, 100 °C}]{\text{KOH,DMSO}}$

DB18C6 = Dibenzo-[18]crown-6

X X

n-C$_3$H$_7$

$\longrightarrow [\quad] \longrightarrow$

X = Cl, Br

The dehydrohalogenation of monohalogenated cyclopropanes under phase transfer conditions has been of greater preparative importance: Monoalkylated bromo-cyclopropanes have been converted to alkylidene cyclopropanes, and 3,3-disubstituted cyclopropylhalides provide the corresponding cyclopropenes in good yield [158].

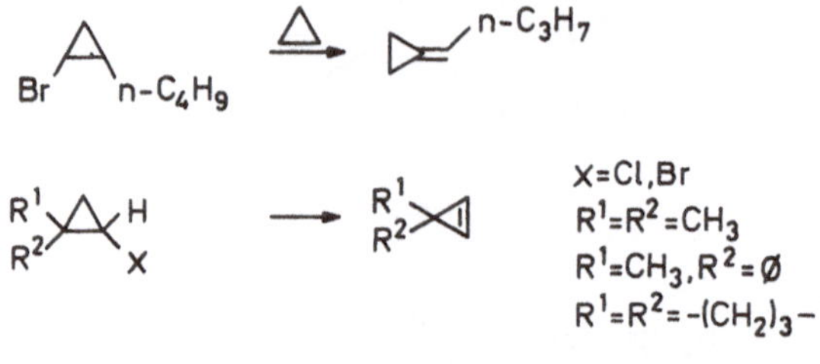

5 Halogen Nucleophilic Substitution (Solvolysis)

As already mentioned, treatment of dihalocyclopropanes with bases furnishes cyclopropenes. When nucleophilic reagents are present, these are added to the strained double bond, and the products thus formed correspond to the products of direct nucleophilic substitution of the substrate, i.e. the elimination/addition process is equivalent to an overall substitution. In fact, in some cases the instable chlorocyclopropene intermediates could be trapped as Diels-Alder adducts with cyclopentadiene.

ref. 159

ref. 159
ref. 160

ref. 161

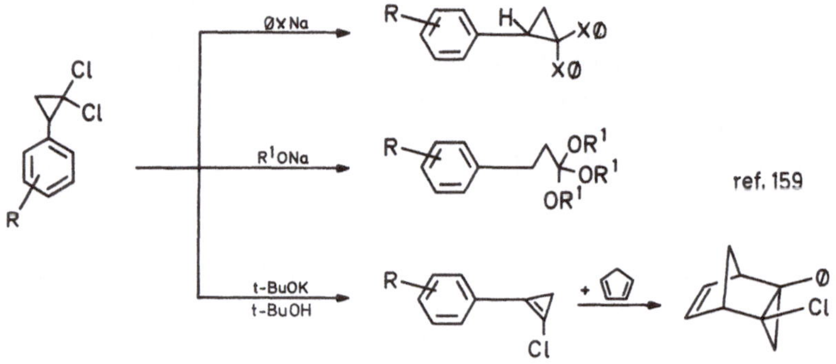

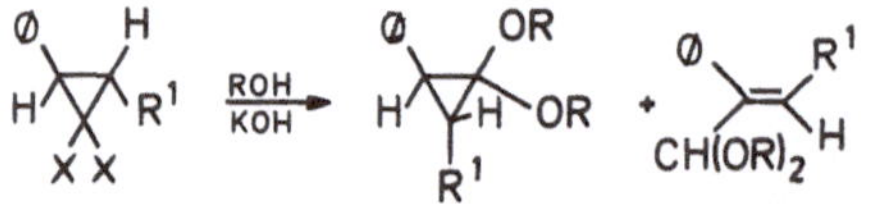

A similar mechanism is thought to be involved when functionalized dihalocyclopropanes are treated with sodium alkoxides as well as thiophenolates or subjected to the influence of potassium hydroxide in methanol.

ref. 162

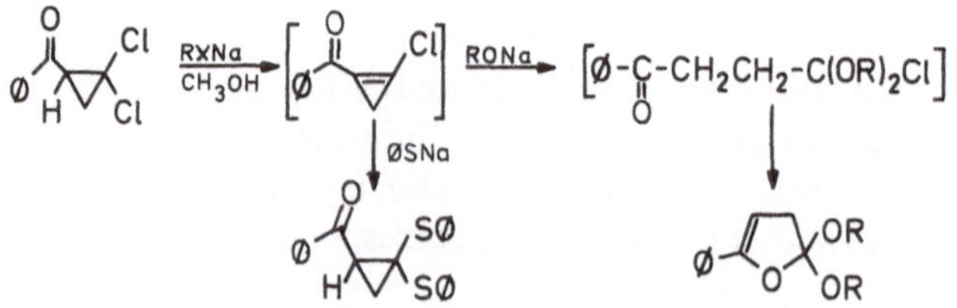

But even if strongly basic reagents are absent, a nucleophilic substitution can take place in *gem*-dihalocyclopropanes. In this case a cyclopropyl-allyl ring-opening occurs, which obeys — as has been known for several years now [164] — the Woodward-Hoffmann orbital symmetry conservation rules. During the ring-opening process the substituents oriented *trans* to the leaving group experience an outward, disrotatory movement. Or, to put it another way, the cyclopropane bond which is opened during the isomerization, attacks the carbon atom which carries the leaving group from the reverse side of the latter.

The allyl cation thus formed may stabilize itself either by readdition of the leaving group — leading to a 2,3-dihalopropene — or by the addition of a nucleophile. The influences of steric and electronic effects on the stereochemistry and on the solvolysis rates of various alkyl-substituted chlorocyclopropanes have been investigated by Parham and co-workers [165, 166], who could show for example that *cis*-2,3-dipropyl-1,1-dichlorocyclopropane solvolyzes 24 times faster than its *trans*-isomer, in accordance with predictions based on orbital symmetry arguments. When one propyl substituent of the *trans*-isomer is replaced by an ethoxy group the rate of solvolysis increases 200 fold.

It is, however, not always possible to draw conclusions on the solvolysis mechanism from the configuration of the products, since the latter may not necessarily be the initially formed solvolysis products. Schlosser has shown for example, that the same product mixture, consisting out of the 2-fluoropropene derivatives VIII–X is obtained from the fluorochlorocyclopropanes shown, regardless whether the (*E*)- or the (*Z*)-isomer is solvolyzed. To rationalize this observation it has been proposed that the tertiary alcohol IX is formed first, and that this intermediate is converted via XI or XII to the more stable primary alcohols VIII and X, respectively [167, 168].

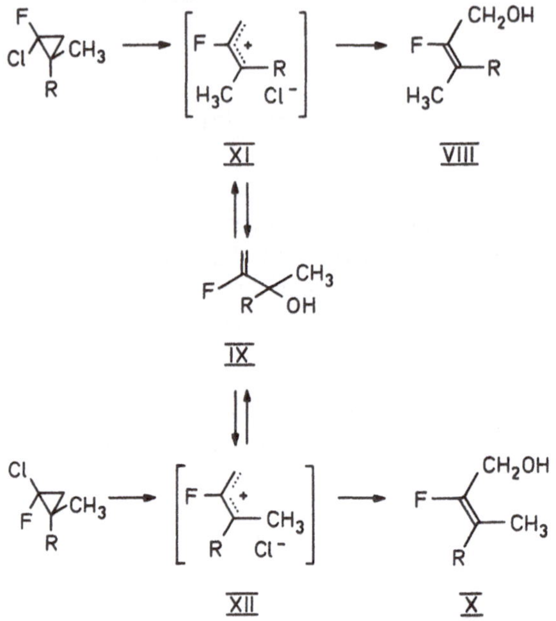

When dihalo-bicyclo[n.1.0]alkanes are solvolyzed, ring-enlargement products are formed. In fact, this reaction is often the method of choice for the preparation of medium-sized rings. As far as the stereochemical outcome of the reaction is concerned, there are conflicting reports in the literature. Thus a mixture of (*E*)- and (*Z*)-2-bromocyclononenols has been obtained by solvolysis of 9,9-dibromo-bicyclo[6.1.0]nonane [169]; in earlier reports only the (*Z*)-diastereomer was mentioned [170], and the solvolysis in aqueous acetone in the presence of silver tosylate apparently provides only the (*E*)-isomer [171].

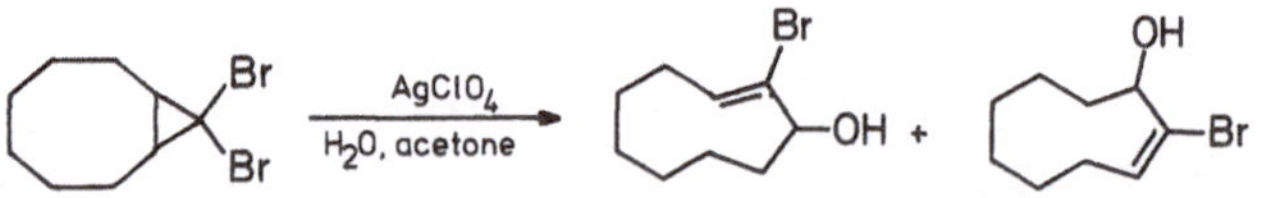

The solvolysis of 7,7-dibromo-bicyclo[4.1.0]heptane in methanol in the presence of silver perchlorate leads to (*E*)-2-bromo-3-methoxycycloheptene [172, 173]. A more detailed study of this reaction — having been performed by Ito in 1986 [174] — has revealed that the stereochemical result of the solvolysis depends on the length of the polymethylene chain in these bicyclic dihalides. For the lower homologs (n = 2 to 4) it is the (*E*)-isomer XIII that is produced, whereas for the more extended systems (n = 5 to 8) the *Z*-bromoethers XIV are produced.

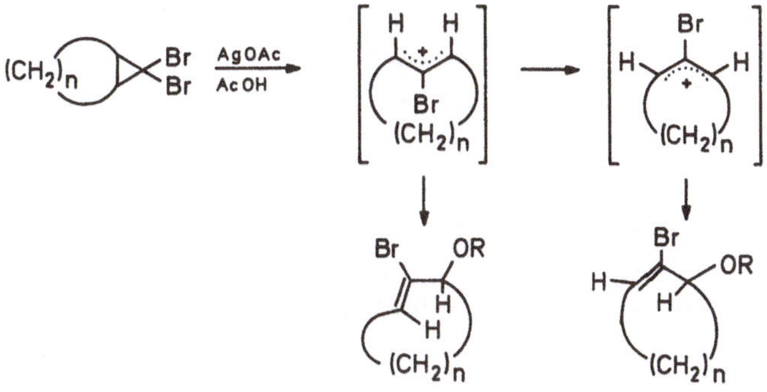

These results correlate with the thermodynamic stabilities of the cycloolefins: rings with n > 10 are more stable if their double bond is *trans*-configurated. Their formation may be understood if free cations are postulated as reaction intermediates which can either isomerize from (Z) to (E) or react with a nucleophile. According to the authors the rate of interception of the various cations by acetate anion cannot compete with the isomerization rate.

Analogous ionic ring-openings have been described for polycyclic *gem*-dibromocyclopropanes. For example, when solvolyzed in 50% aqueous acetone in the presence of triethylamine, the already mentioned dibromocarbene adduct of benzvalene loses its *endo*-bromine substituent and opens in a disrotatory fashion to a tricycloheptenyl cation which may be intercepted by water to provide the bromoalcohol shown [175].

The carbocation intermediate, incidentally, is an isomer of the tropylium ion.

Ionic ring-opening reactions of cyclopropanes under solvolytic conditions followed by ring-closure of the primarily produced cations have been employed for the synthesis of lactones, tetrahydrofuranes, and pyranes, respectively, as shown below [176].

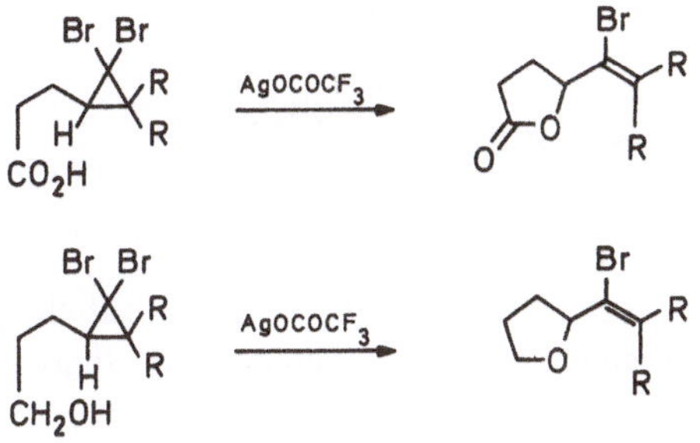

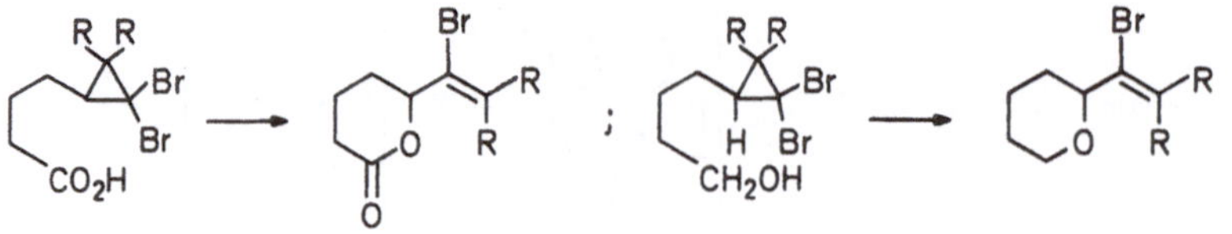

Products with a [5]metacyclophane skeleton are obtained when dihalocarbene adducts of certain cycloheptene derivatives are treated with silver(I)perchlorate in the presence of 2,4,6-collidine in tetrahydrofuran [177, 178].

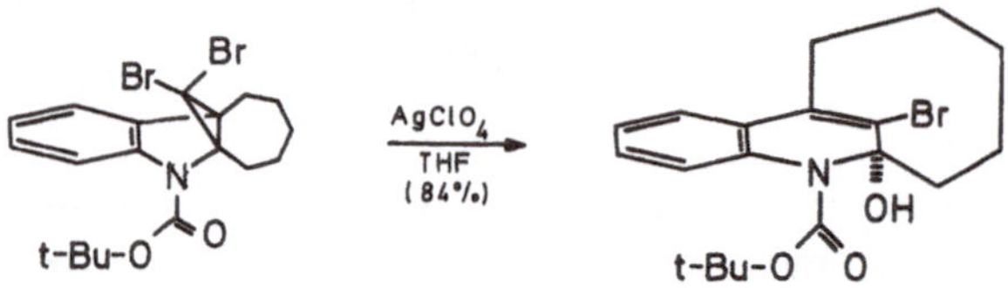

A number of *gem*-dibromocyclopropane derivatives has been reacted with aromatic compounds in the presence of aluminum chloride or ferric chloride providing indenes in yields up to 80%. To rationalize this interesting anellation process it has been proposed that the cyclopropyl cation formed under the influence of the Lewis acid collapses to an allylic ion, which then functions as the alkylating agent [179].

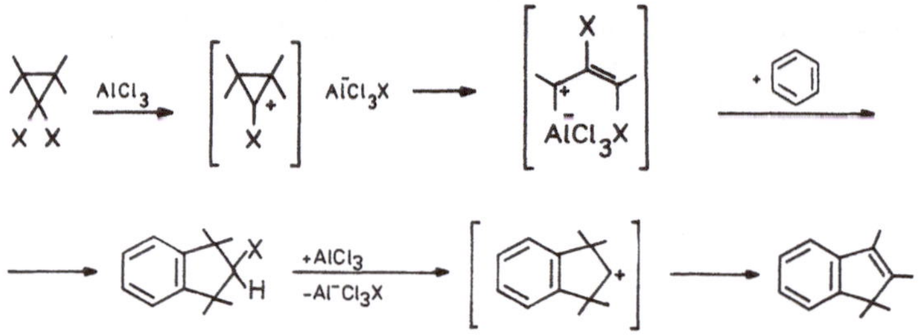

Allylic cations have also been discussed as intermediates in the conversion of *gem*-dichlorocyclopropylethers in alcoholic solvents in the presence of base resulting in the formation of acetals of the general structure $RR^1C=CClCH-(OEt)OR^2$ in good yields [180, 181].

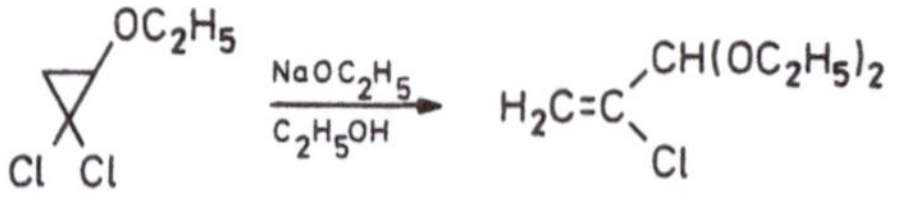

A silver-induced ring-opening reaction of various tri- and tetrahalocyclopropanes to produce polyhalogenated dienes or methyl polyhaloallyl ethers in good yields has recently been described [182]:

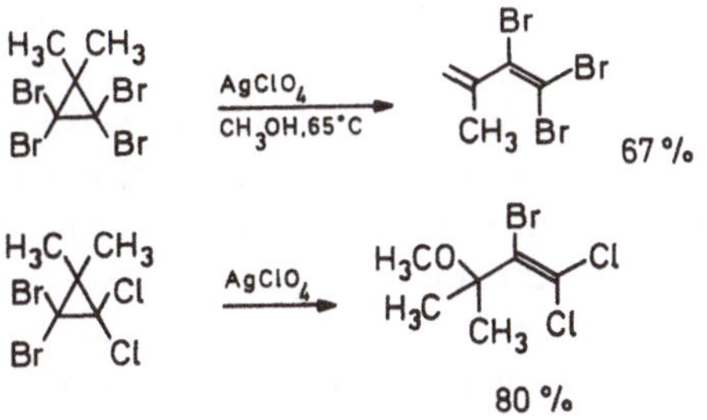

Some rare examples are known in which a cyclopropane ring is preserved during solvolysis, i.e. its halogen atoms are substituted by a nucleophilic group. When the substituted bicyclic substrates XV are solvolyzed, ring-enlarged products are formed exclusively for R=Cl, OCH$_3$, and CH$_3$. However, with R being trimethylsilyl the three-membered ring remains intact [184]:

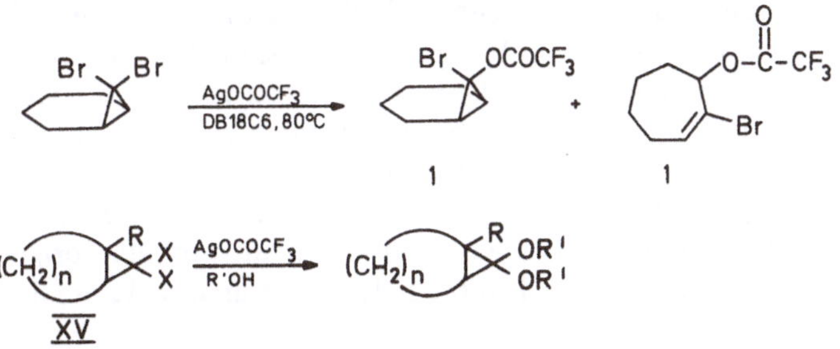

6 Thermal Rearrangements

One of the classical papers in this area is Neureiter's discovery of the vinyl-cyclopropane to cyclopentene rearrangement [185]. This important process was first observed on a 1,1-dichloro-2-vinylcyclopropane derivative, which on heating yielded a chlorocyclopentadiene, presumably by the loss of hydrochloric acid from the initially formed 4,4-dichlorocyclopentene.

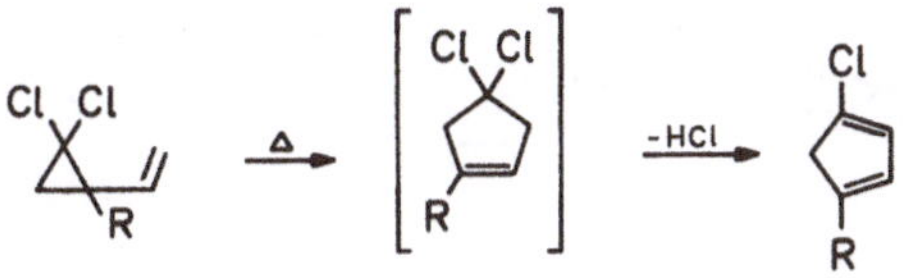

Since then numerous thermal rearrangements of dihalocyclopropanes have been observed and the isomerization has been reviewed several times [11, 186]. It therefore suffices to discuss some special features of these reactions here. The thermal isomerization of dihalocyclopropanes is often accompanied by a halogen migration and then leads to 2,3-dihalogenated propenes.

At first sight, one could suppose that this isomerization follows the Woodward-Hoffmann-allowed cyclopropyl-allyl pathway (see above). At least in some cases, however, the required heterolytic bond rupture is quite unlikely to take place under the reaction conditions (gas phase or apolar solvents). On the other hand, the diradical mechanism which has been invoked to explain the thermal ring-opening of alkyl substituted cyclopropanes, fails to account for certain observations made with halocyclopropanes. For example, radical inhibitors do neither effect the rate of the thermal conversion of 1,1-dichlorocyclopropane to 2,3-dichloro-1-propene nor do they influence the nature of the reaction products [187]. From the activation parameters for the thermolysis of *cis*-1,1-dibromo-2,3-dimethylcyclopropane it has been concluded that halogen migration and ring-rupture occur in a concerted fashion via a cyclic transition state [188].

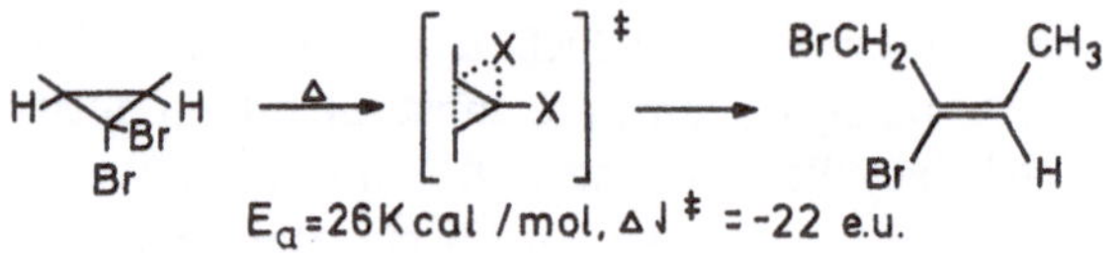

$$E_a = 26 \text{ Kcal/mol}, \Delta\downarrow^{\ddagger} = -22 \text{ e.u.}$$

Other authors have postulated similar transition states for these isomerizations [189, 190]. According to extended-Hückel calculations the migration of an *endo*-halogen atom in certain bicyclic systems should be accompanied by a conrotatory ring-opening [191]. Thermal isomerizations of bi- and tricyclic substrates incorporating a halocyclopropane moiety have been studied in some detail. In these reactions the endocyclic bond of the three-membered ring (i.e. the zero bridge) is broken and cyclic dihaloolefins or consecutive products result. The ease, with which these reactions often occur, probably originates from the strain release which occurs when the polycyclic frameworks are destroyed.

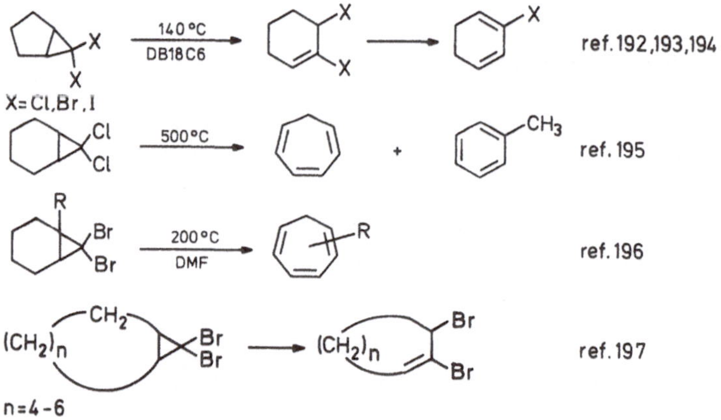

The *endo*-adducts of dihalocarbenes onto norbornadiene and norbornene are thermally so labile that they even isomerize under the conditions of their formation.

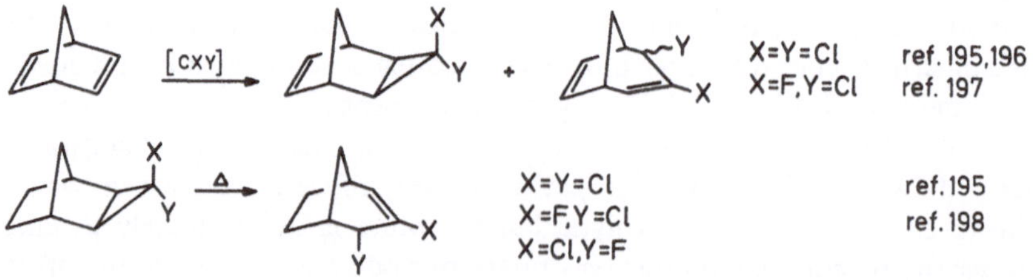

A detailed discussion of the thermal isomerization of *gem*-dihalocyclopropane — including polycyclic substrates — has been presented by Ioffe and Nefedov. It is proposed (and theoretically justified by MO calculations) that these concerted rearrangements of *gem*-dihalocyclopropanes to 2,3-dihaloolefins take place by synchronous migration of the halogen and rupture of the opposite bond of the cyclopropane ring [191]. The dibromocarbene adduct of benzvalene isomerizes in inert solvents like carbon tetrachloride to 4,5-dibromotricyclo[4.1.0.0^{2,7}]hept-3-ene when heated to 80 °C. Ring-opening presumably takes place by removal of the *endo*-bromine substituent and disrotatory opening of the C^1—C^6-bond of the starting *gem*-dihalide [199].

The thermal isomerization of dihalocyclopropanes is facilitated by the presence of aryl, alkyl and alkoxy substituents in the three-membered ring. Phenyl substituted systems may give rise to indenes, and in the alkyl series the dehydrohalogenation of 2,3-dihalocyclopropane derivatives may lead to dienes.

ref. 200

ref. 200

ref. 201

$R^1, R^2 = H, CH_3, C_2H_5, n\text{-}C_3H_7 ; R = CH_3, C_2H_5$

The yields of these thermal processes are low, however, especially if they are employed for the preparation of a specific target molecule. For example, the pyrolysis of 1,1-dichloro-2,2-dimethyl-cyclopropane furnishes a product mixture (total yield 63%) which contains the diene XVI in 24% yield only.

Schlosser and co-workers have developed a method for the preparation of fluoroisoprenes with yields ranging between 38 and 80%, consisting of the dehydrohalogenation of 2-chloro-2-fluoro-1-halomethylcyclopropanes at 100 °C in diglyme in the presence of zinc [203] or under the influence of potassium hydride [204].

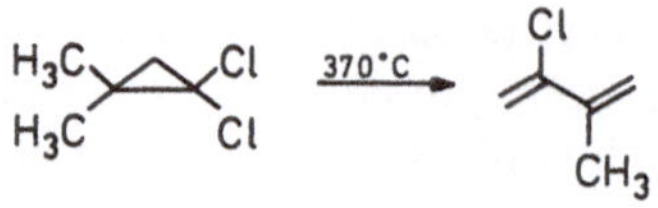

ref. 202

XVI

R=H,R^1=CH$_3$; R=CH$_3$,R^1=H
X=Cl,Br, I

When dihalocyclopropanes carrying a cyclopropyl substituent are thermolyzed both rings are opened and substituted 1,3-hexadienes are formed.

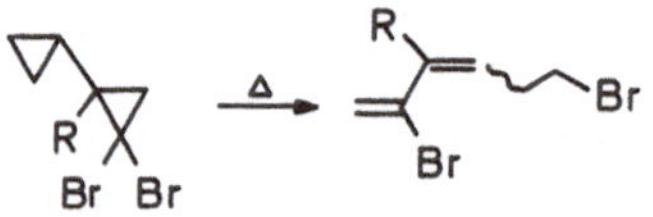

R=ferrocenyl   ref. 205

R=cyclopropyl   ref. 206,207

The thermal isomerization of 2-aryl-1,1-dicyclopropyl-3,3-dihalocyclopropanes has been interpreted as a $[\sigma_2 + \sigma_2 + \sigma_2]$process in which three σ-bonds are broken in a concerted fashion [208].

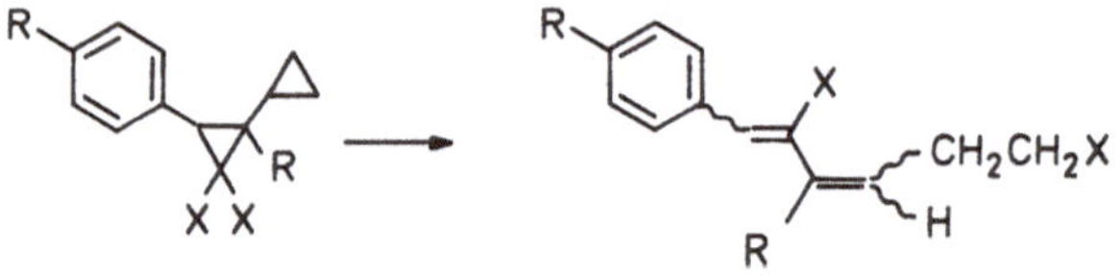

R=H,Cl,CH$_3$ R=CH$_3$,cyclopropyl
X=Cl,Br

6.1 Vinylcyclopropane to Cyclopentene Isomerizations

The thermal rearrangement of vinylcyclopropanes to cyclopentenes is one of the most thoroughly studied thermal processes known. It constitutes an important route to five-membered ring-systems, and has been applied widely in synthesis, even for the preparation of structurally complex products, i.e. numerous natural products [209]. This ring-enlargement has also been observed with dihalocyclopropanes, providing the expected dihalocyclopentenes. Like the parent system the halogen substituted vinylcyclopropanes isomerize via diradical intermediates [210]. The rearrangement rate for the latter is, however, much higher than that for the halogen-free substrates.

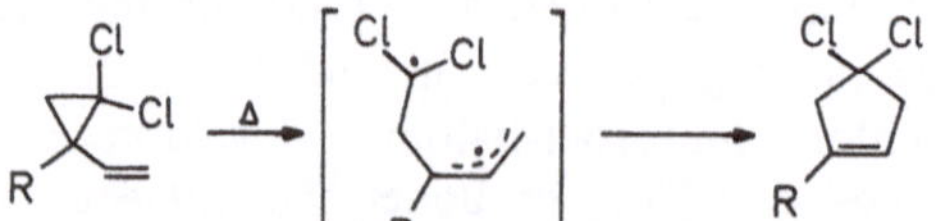

R=H,CH$_3$

Whereas the parent difluoro-vinylcyclopropane isomerizes to difluorocyclopentene under pyrolysis conditions, the corresponding alkyl compounds also lead to acyclic dienes. The activation energy for the difluoro-vinylcyclopropane isomerization is practically identical with that observed for the unsubstituted hydrocarbon [211, 212]. If the alkyl group is oriented *cis* to the vinyl substituent, only dienes are isolated, and the process occurs at much lower temperatures. Presumably these stereoisomers rearrange by a different mechanism (a 1,5-homodienyl hydrogen shift [213]). When the dichlorocyclopropane XVII is subjected to flash vacuum pyrolysis it isomerizes to 9,9-dichloro-bicyclo[5.3.0]dec-1(7)-ene [214].

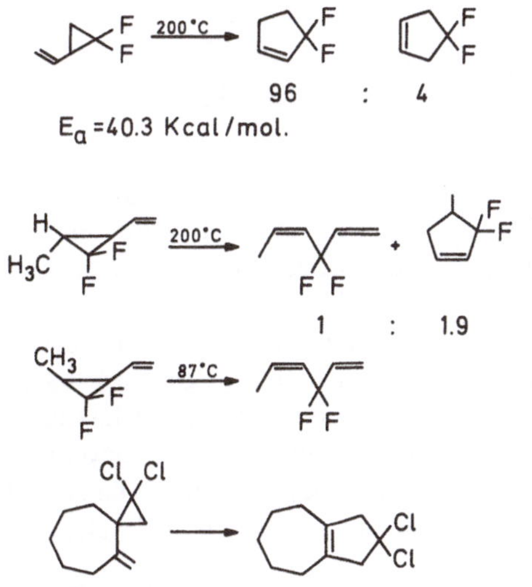

$E_a = 40.3$ Kcal/mol.

An interesting 1,2-divinylcyclopropane isomerization is observed when 5,5,10,10-tetrachlorotricyclo[7.1.0.0^{9,6}]deca-2,7-diene is subjected to flash vacuum pyrolysis: at 700 °C and 10^{-4} Torr a mixture of three isomeric dichloroazulenes is produced [215].

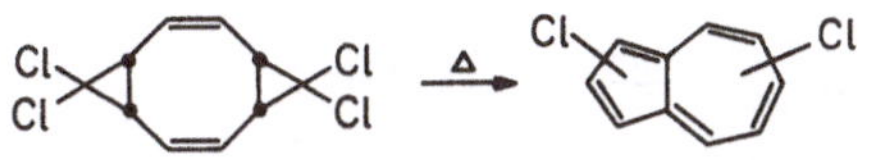

6.2 Thermal Rearrangements of Dichloromethylene-Cyclopropanes

At elevated temperatures, methylenecyclopropane and its derivatives undergo a rearrangement which maintains the methylenecyclopropane skeleton. Obviously, for the parent system this process is degenerate [216, 217–222]. The 2,2-dihalo-methylene-cyclopropanes behave analogously, providing either mixtures of the methylenecyclopropanes XVIII and XIX or only the dichloride XVIII, with the actual outcome of the reaction depending on the substituents again.

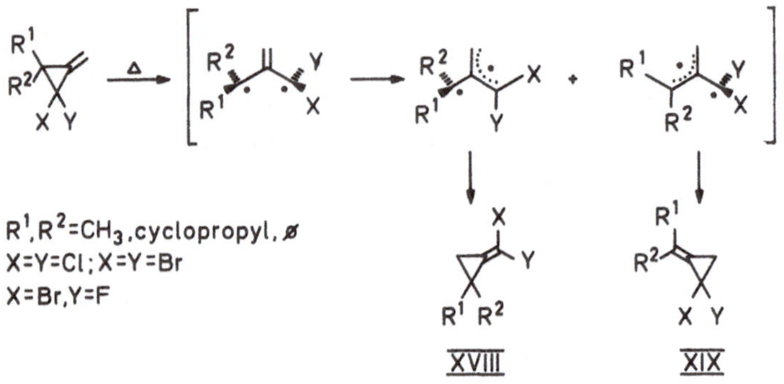

$R^1, R^2 = CH_3$, cyclopropyl, $\emptyset$
$X = Y = Cl; X = Y = Br$
$X = Br, Y = F$

Chlorine and bromine as well as cyclopropyl and phenyl groups increase the rate of isomerization significantly by lowering the activation enthalpy. With an activation entropy approaching zero the process most likely occurs via diradicals [223].

Among the other reactions of the *gem*-dihalomethylene cyclopropanes the cycloadditions are of greatest importance, although they have so far been observed only for the fluorine containing compounds. It has been shown, that 2,2-difluoro-1-methylene-cyclopropane prefers the [4+2]cycloaddition mode, whereas the isomeric 1-(difluoromethylene)cyclopropane behaves like a typical fluoroolefin, i.e. it is a poor dienophile but active in [2+2]cycloadditions [224]:

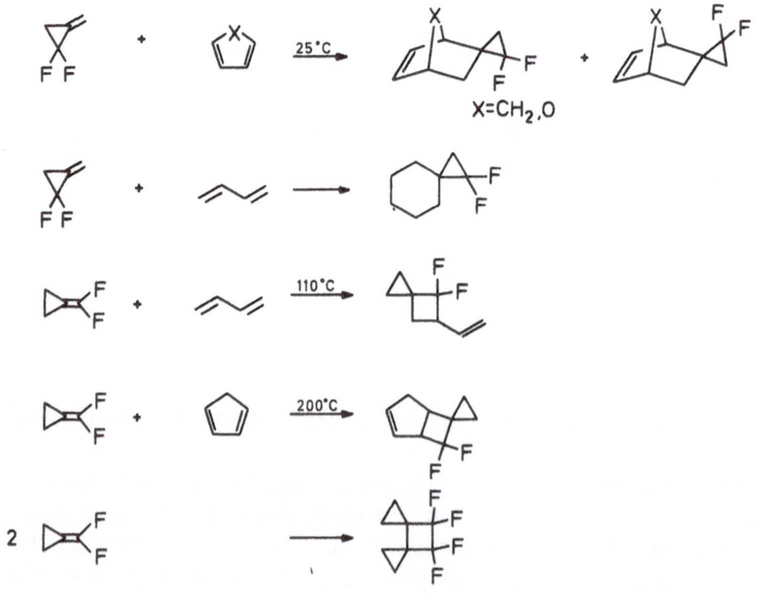

6.3 Dihalocyclopropanes as Unstable Intermediates

Although the *gem*-dihalocyclopropanes are fairly stable compounds, they can participate — as has been shown in the above sections — in quite a number of chemical transformations. Several reactions between dihalocarbenes and alkenes have been described in which no dihalocyclopropane formation could be observed; that these intermediates might have been produced was only inferred from the type of products finally isolated. A typical process of this type is the *endo*-addition of dihalocarbenes to norbornene and norbornadiene as discussed above. Comparable rearrangements have been observed, when dichlorocarbene additions either lead to aromatic products or when they cycloadd to rather inert aromatic systems. In the latter case a ring-enlargement takes place. A reaction related to the concerted opening of two cyclopropane rings in a bicyclopropyl system as discussed above takes place when dichlorocarbene is added to spiro[2.4]hepta-4,6-diene [227].

ref. 225

ref. 226

ref. 227

When pyrroles, triazoles and other nitrogen containing heterocyclic compounds react with dichlorocarbene, no addition to their carbon-carbon double bonds has been observed. In some cases the formation of cyclopropane adducts has been suspected; however, alternate ways to explain the reaction products cannot be excluded [228].

7 Conclusion and Outlook

To summarize, *gem*-dihalocyclopropanes may serve as starting materials for the preparation of cyclopropane and cyclopropene derivatives, they can lead to compounds with bicyclobutane and spiropentane structures, provide allenes and higher cumulenes, cyclopentenes and cyclopentadienes, and many other classes of compounds.

A few final examples may illustrate the quite general and steadily growing use of these versatile three-membered ring compounds.

Dihalocyclopropanes have been employed in the crucial steps of a synthesis of crepinine, chrysanthemic acid and (±)-pyrenoforin. Altogether the large variety of reaction types and ease of access to the dihalocyclopropanes makes them very useful in organic synthesis. Especially the possibility of introducing other functional groups as replacement for the halogen atoms makes them attractive substrates for the preparation of a large variety of practically important compounds like pheromones, prostaglandins and various insecticides [232].

ref. 229

ref. 230

ref. 231

On the other hand there are certain groups of reactions that have been neglected so far, among them the photochemical behavior of the *gem*-dihalocyclopropanes. One of the few reported examples is by Weyerstahl who has attempted to generate hexachloro-tris-σ-homobenzene from the corresponding tris-σ-homocyclo-heptatrienone [233]. The latter, however, prefers to undergo *trans-cis* isomerization rather than decarbonylation.

It is not unreasonable to expect preparatively useful photoreactions of appropriately functionalized dihalocyclopropanes also. Although the present trend in *gem*-dihalocyclopropane chemistry clearly favors the use of more highly functionalized substrates, the traditional role — mentioned in the introduction — of employing these strained molecules in hydrocarbon synthesis, is by no means an approach of the past. This point is stressed by the recent synthesis of octavalene by Christl and his students who used halocarbene adducts during several crucial transformations in their preparation of this interesting C_8H_8-hydrocarbon [234 to 237].

Whereas the ring-enlargement of benzvalene to homobenzvalene and the second dibromocarbene addition posed no problems, the final insertion step provided 5-ethinyl-1,3-cyclohexadiene instead of the expected octavalene (tricyclo[$5.1.0.0^{2,8}$]-octa-3,5-diene). However, when homobenzvalene is first reacted with mono-bromocarbene and the resulting adduct with pyridinium perbromide, a tribromide is obtained, which after thermal ring-opening and dehydrobromination is finally converted to the target molecule by treatment with *tert*-butyllithium. In this sequence, two carbon atoms have eventually been inserted between the carbon atoms of the double bond which bridged the bicyclobutane moiety of the original molecule [238].

8 References

1. Gustavson G (1890) J. Prakt. Chem., Neue Folge 42: 495
2. Doering WvE, Hoffmann AK (1954) J. Am. Chem. Soc. 76: 6162
3. One of the numerous, more recent applications of this methodology — leading to novel polyspiranes — has been described by Fukuda Y, Yamamoto Y, Kimura K, Odaira Y (1979) Tetrahedron Lett. 877
4. Trost BM (1986) Topics Curr. Chem. 133: 5; cf. de Meijere A (1979) Angew. Chem. 91: 867; (1979) Angew. Chem. Int. Ed. Engl. 18: 809
5. Chinoporos E (1973) Chem. Rev. 63: 235
6. Parham WE, Schweizer EE (1963) Org. Reactions 13: 55
7. Jerosch-Herold B, Gaspar PP (1965) Fortschr. Chem. Forsch. 5: 89
8. Kirmse W (1971) Carbene chemistry, 2nd ed, Academic, New York
9. Closs GL In: Hart H, Karabatsos GJ (eds) Advances in alicyclic chemistry, vol 1 Academic, New York, p 53
10. Wendisch D (1971) In: Houben-Weyl-Müller, Methoden der organischen Chemie, vol IV/3, Thieme, Stuttgart; A new volume of the Houben-Weyl series has recently been published dealing with divalent carbon species: Regitz M, Hanack M (eds) (1989) Vol E 19, Niedervalente Kohlenstoffverbindungen Thieme, Stuttgart

Rafael R. Kostikov et al.

11. Weyerstahl P (1983) In: Patai S (ed) The chemistry of functional groups, The chemistry of halides, pseudo-halides and azides, part 2, Wiley, New York, p 1451
12. Zefirov NS, Kazimirchik IV, Lukin KA (1985) The cycloaddition of dichlorocarbene to olefins, Nauka, Moscow, p 151
13. Keller WE (1986) Phase-transfer reactions, Thieme, Stuttgart, p 106
14. Haddon RC, Chichester SV, Stein SM, Marshall JH, Mujsce AM (1987) J. Org. Chem. 52: 711
15. Liebowitz SM, Johnson HJ (1986) Synth. Commun. 16: 1255
16. Krief A, Laboureur JL, Dumont W (1987) Tetrahedron Lett. 28: 1549
17. Perrett AT, Laurie VW (1975) J. Chem. Phys. 62: 2469
18. Dolbier WR, Piedrahita CA, Al-Sader BH (1979) Tetrahedron Lett. 2957
19. Dolbier WR, Odaniecz M, Gomulka E, Saskolski M, Koroniak H (1984) Tetrahedron 40: 3945
20. Flygare WH, Narath A, Gwinn WD (1962) J. Chem. Phys. 36: 200
21. Cole KC, Gilson DFR (1975) J. Mol. Struct. 28: 385
22. Alekseev NV, Bardsjan AP, Shostakowsky VM (1972) Zh. Struct. Khim. 13: 512
23. Skancke A, Flood E, Boggs J (1977) J. Mol. Struct. 40: 263
24. Skancke A (1977) J. Mol. Struct. 42: 235
25. Isogai K, Nishizawa N, Saito T, Sakai J (1983) Bull. Soc. Chem. Japan 56: 1556
26. Nefedov OM, Shafran RN (1965) Izv. Akad. Nauk, Ser. Khim. 538
27. Nefedov OM, Shafran RN, Novizkaya NN (1972) Zh. Org. Khim., 2075
28. Schuster DI, Lee FF (1965) Tetrahedron Lett. 4119
29. Oku A, Tsuji H, Yoshida M, Yoshiura N (1981) J. Am. Chem. Soc., 103: 1244
30. Oku A, Narada K, Yagi T, Shirahose Y (1983) J. Am. Chem. Soc., 105: 4400
31. Nefedov OM, Agaveljan AS (1974) Izv. Akad. Nauk. SSR, Ser. Khim 838
32. Mehta G, Kapoor SK (1974) J. Organomet. Chem. 80: 213
33. Jakushkina NI, Zacharova GA, Surmina LS, Bolesov IG (1980) Zh. Org. Khim. 16: 1834; (1980) J. Org. Chem. USSR 16: 1551
34. Djachenko AI, Korneva OS, Nefedov OM (1980) Izv. Akad. Nauk SSR, Ser. Khim 2432
35. Djachenko AI, Korneva OS, Nefedov OM (1982) Izv. Akad. Nauk SSR, Ser Khim. 2842
36. At elevated temperatures, e.g. in a high-boiling solvent, LiAlH$_4$ can be used to fully reduce gem-dichlorocyclopropanes: Kaufmann D, Fick H-H, Schallner O, Spielmann W, Meyer L-U, Gölitz P, de Meijere A (1983) Chem. Ber. 116: 587
37. Jakushkina NI, Bolesov IG (1979) Zh. Organ. Khim. 15: 954; (1979) J. Org. Chem. USSR 15: 853. The use of tri-n-butylhydride as a reducing reagent has been reviewed recently: Neumann WP (1987) Synthesis 665. This summary includes a section on the selective reduction of gem-dihalocyclopropanes
38. Sydnes LK, Skattebøl L (1974) Tetrahedron Lett. 3703
39. Sydnes LK, Skattebøl L (1978) Acta Chem. Scand. B32: 632
40. Mirao T, Masunaga T, Ohshiro Y, Agawa T (1981) J. Org. Chem. 46: 3745
41. Djachenko AI, Korneva OS, Abramova NM, Nefedov OM (1984) Izv. Akad. Nauk SSR, Ser. Khim. 2818
42. Djachenko AI, Rudashewskaja TY, Korneva OS, Shteinshneider AY, Nefedov OM (1978) Izv. Akad. Nauk SSR, Ser. Khim. 2191
43. Djachenko AI, Korneva OS, Nefedov OM (1984) Izv. Akad. Nauk SSR, Ser. Khim. 2673
44. Jefford CW, Burger U, Laffer MH, Kabengele T (1973) Tetrahedron Lett. 2483
45. Christl M, Freitag G, Brüntrup G (1978) Chem. Ber. 111: 2307
46. Gaoni Y (1981) J. Org. Chem. 46: 4502
47. Kulinkovitch OG, Tishenko IG, Romashin JN (1984) Izv. Akad. Nauk SSR, Ser. Khim. 20: 1422
48a. Meijs GF (1987) J. Org. Chem. 52: 3923; 48b. Meijs GF, Doyle IR (1985) J. Org. Chem. 50: 3713; cf. Oshima K, Shirajuji T, Yamamoto I, Nozaki H (1973) Bull. Chem. Soc. Japan 46: 1233
49. Ishihara T, Ontani E, Ando T (1975) J. Chem. Soc. Chem. Commun. 367
50. Yamanaka H, Yagi T, Teramura K (1971) J. Chem. Soc., Chem. Commun. 380

51. Sydnes L, Skattebøl L (1975) Tetrahedron Lett. 4603; cf. Sydnes L, Skattebøl L (1978) Acta Chem. Scand. B32: 632
52. Santelli C (1980) Tetrahedron Lett. 21: 2893
53. Barlett R (1978) J. Org. Chem. 43: 3500
54. Jorgensen E, Sydnes LK (1986) J. Org. Chem. 51: 1926
55. Kitatani K, Hijama T, Nozaki H (1975) J. Am. Chem. Soc. 97: 949
56. Hijama T, Kanakura A, Morizawa Y, Nozaki H (1982) Tetrahedron Lett. 23: 1279
57. Fermanovsky AA, Kosicina N, Bolesov IG (1981) Zh. Org. Khim. 17: 1778
58. Warner PM, Herold PD (1983) J. Org. Chem. 48: 5411
59. Warner PM, Chang S-C, Roszewski NJ (1985) Tetrahedron Lett. 26: 5371
60. Sander V, Weyerstahl P (1976) Angew. Chem. 88: 259; (1976) Angew. Chem. Int. Ed. Eng. 15: 244
61. Norden W, Sander V, Weyerstahl P (1983) Chem. Ber. 116: 3097
62. Kulinkovitch OG, Tishenko IG, Sviridov SV (1986) Zh. Organ. Khim. 22: 1416
63. Hijama T, Takehara S, Kitatani K, Nozaki H (1974) Tetrahedron Lett. 3295
64. Kulinkovitch OG, Tishchenko IG, Sviridov SV (1986) Zh. Organ. Khim. 22: 1416; (1986) J. Org. Chem. USSR 22: 1275
65. Braun M, Dammann R, Seebach D (1975) Chem. Ber. 108: 2368
66. Hopf H (1980) In: Patai S (ed) The chemistry of ketenes, allenes and related compounds, part 2, Wiley, New York p 779
67. Skattebøl L (1961) Tetrahedron Lett. 167
68. Skattebøl L (1963) Acta Chem. Scand. 17: 1683
69. Schuster HE, Coppola GM (1984) Allenes in organic synthesis, Wiley-Interscience, New York
70. Landor SR (ed) (1982) The chemistry of the allenes vols I–III, Academic, London; cf. Smadja W (1983) Chem. Rev., 83: 263
71. Pasto DJ (1984) Tetrahedron 40: 2805
72. Besides the alkyllithium compounds numerous other reagents have been used to initiate the dehalogenation, including such systems as $CrCl_3/LiAlH_4$ (Okuda Y, Hiyama T, Nozaki H (1977)) Tetrahedron Lett. 3829; and Cu(O)/isonitrile complexes (Crozet MP, Surzur JM, Jauffred R, Ghiglione C (1979)) Tetrahedron Lett. 3077. A more recent — and quite promising — method employs lithium or magnesium in tetrahydrofuran under ultrasonic irradiation (Yu T (1985)) Tetrahedron Lett. 26: 4231. — Especially for the preparation of conjugated and non-conjugated bisallenes the DMS route is the method of choice and very often the only way to prepare these interesting tetraenes in acceptable yields. When appropriately substituted, conjugated bisallenes may be either *meso-* or *d,l*-compounds (Kleveland K, Skattebøl L (1975) Acta Chem. Scand. B29: 827 and references cited therein; cf. Becher G, Skattebøl L (1979) Tetrahedron Lett. 1261
73. Roth WR, Schmidt T, Humbert H (1975) Chem. Ber. 108: 2171
74. Dewar MJS, Fonken GJ, Kirschner S, Minter SE (1975) J. Am. Chem. Soc. 97: 6750
75. Jakushkina NI, Bolesov IG (1979) Zh. Org. Khim 15: 311; (1979) J. Org. Chem. USSR 15: 270; cf. Slobodin YM, Egenburg IZ, Khachaturov AS (1974) Zh. Org. Khim. 10: 21; (1974) J. Org. Chem. USSR 10: 18; Kiselev MY, Kostikov RR, Molchanov AP (1989) Zh. Org. Khim 25: 870
76. Kostikov RR, Molchanov AP, Nagi SM (1983) Zh. Org. Khim. 19: 1437; (1983) J. Org. Chem. USSR 19: 1291; cf. Nefedov OM, Dolgii IE, Bulusheva EV (1978) Bull. Acad. Sc. USSR 27: 1271; Skattebøl L (1963) Tetrahedron Lett. 2175
77. Bee LK, Beeby J, Everett JW, Garratt PJ (1975) J. Org. Chem. 40: 2212; cf. Suvorova GN, Komendantov MI (1979) Zh. Organ. Khim. 15: 1435; (1979) J. Org. Chem. USSR 15: 1280
78. Blickle P, Hopf H (1978) Tetrahedron Lett. 449
79. Heldeweg RF, Hogeveen H (1978) J. Org. Chem. 43: 1916
80. Carlton JB, Levin RH (1976) Tetrahedron Lett. 3761
81. Melane RC, Schuster GB (1983) J. Org. Chem. 48: 810
82. Moore WR, Ward HR, Merritt RF (1961) J. Am. Chem. Soc. 83: 2019
83. Paquette LA, Zon G (1974) J. Am. Chem. Soc. 96: 203

84. Paquette LA, Browne AR, Chamont E, Blount JF (1980) J. Am. Chem. Soc. 102: 643; cf. Vogel E, Wassen J, Königshofen H, Müllen K, Oth JFM (1974) Angew. Chem. 86: 777; (1974) Angew. Chem. Int. Ed. Engl. 13: 732; Vogel E, Kerimis D, Allison NT, Zellerhoff R, Wassen J (1979) Angew. Chem. 91: 579, (1979) Angew. Chem. Int. Ed. Engl. 18: 545
85. Warner P, Chang S-C (1979) Tetrahedron Lett. 7141 and literature cited
86. Carlton JB, Levin RH, Clardy J (1976) J. Am. Chem. Soc. 98: 6068; cf. Ref. 80
87. Harnos S, Tivakornpannarei S, Waali EE (1986) Tetrahedron Lett. 27: 3701; cf. Moore WR (1972) J. Am. Chem. Soc. 94: 4753; Moore WR (1972) J. Am. Chem. Soc. 94: 4753; — For novel uses of 1,2-cyclohexadiene in organic synthesis see Christl M, Schreck M (1987) Angew. Chem. 99: 474, (1987) Angew. Chem. Int. Ed. Engl. 26: 449; Christl M, Schreck M (1987) Chem. Ber. 120: 915
88. Price JD, Johnson RP (1986) Tetrahedron Lett. 4679; cf. Gardner PD (1066) Tetrahedron Lett. 2793; Theoretical calculations concerning the stability of cycloallenes have been published by Angus RO, Schmidr MW, Johnson RP (1985) J. Am. Chem. Soc. 107: 532; cf. Johnson RP (1989) Chem. Rev. 89: 1111
89. Schreck M, Christl M (1987) Angew. Chem. 99: 720; (1987) Angew. Chem. Int. Ed. Engl. 26: 690. See also the references in this paper concerning previous reports on the dehalogenation of the 6,6-dihalo-3-oxa-bicyclo[3.10]hexanes
90. Christl M, Lang R, Lechner M (1980) Liebigs Ann. Chem. 980; cf. Christl M, Lechner M (1975) Angew. Chem. 87: 815, (1975) Angew. Chem. Internat. Ed. Engl. 14: 765. For a comprehensive study of the addition of dihalocarbenes to benzvalene see Christl M, Freitag G, Brüntrup G (1978) Chem. Ber. 111: 2307; Christl M, Herzog C, Brückner D, Lang R (1086) Chem. Ber. 119: 141
91. Taylor RT, Paquette LA (1975) Angew. Chem. 87: 488, (1975) Angew. Chem. Internat. Ed. Engl. 14: 496
92. Christl M (1981) Angew. Chem. 93: 515, (1981) Angew. Chem. Int. Ed. Engl. 20: 529
93. Moore WR, Hill JB (1970) Tetrahedron Lett. 4553; cf. Dehmlow EV, Ezimora GC (1978) Tetrahedron Lett 1599
94. Nilsen NO, Skattebøl L, Baird MS, Buxton SR, Slowey PD (1984) Tetrahedron Lett. 2887
95. Wiberg K, Walker FH (1982) J. Am. Chem. Soc. 104: 5239
96. Semmler K, Szeimies G, Belzner J (1985) J. Am. Chem. Soc. 107: 6410
97. Belzner J, Bunz U, Semmler K, Szeimies G, Opitz K, Schlüter AD (1989) Chem. Ber. 122: 397
98. Bunz U, Szeimies G (1989) Tetrahedron Lett. 30: 2087
99a. Moore WR, Ward HR (1960) J. Org. Chem. 25: 2073; 99b. Warner P, Chang S, Powell DR, Jacobson RA (1981) Tetrahedron Lett. 22: 533; 99c. Cf. footnote in Lord RC, Wurrey CJ (1974) Spectrochimica Acta 30A: 915; 99d. Anderson HW (1972) Ph. D. thesis, MIT, Cambridge, Massachusetts; de Meijere A, Egge R (unpublished results); Egge R (1981) Diplomarbeit, Universität Hamburg
100. Köbrich G, Goert W (1968) Tetrahedron 24: 4327
101. Jones M, Petrillo EW (1969) Tetrahedron Lett. 3953; The addition of dichloro- and dibromocarbene to various 1,4-dihydrobenzenes and isotetralin has been investigated extensively by E. Vogel and co-workers, cf. Ippen J, Vogel E (1974) Angew. Chem. 86: 778, 780, (1974) Angew. Chem. Int. Ed. Engl. 13: 734, 736 and references cited therein. — For an interesting use of 7,7-dibromobicyclo[4.1.0]hept-3-ene during the preparation of the naturally occuring trop one nezukone see Banwell MG, Garatt GL, Richard CEF (1985) J. Chem. Soc., Chem. Commun. 514
102. Skattebøl L (1966) J. Org. Chem. 31: 2789
103. Brinker UH, Gomann K, Zorn R (1983) Angew. Chem. 95: 893, (1983) Angew. Chem. Internat. Ed. Engl. 22: 869
104. Brinker UH, Streu J (1980) Angew. Chem. 92: 641, (1980) Angew. Chem. Internat. Edit. Engl. 19: 631
105. Brinker UH, Wüster H, Maas G (1987) 99: 585, (1987) Angew. Chem. Internat. Ed. Engl. 26: 577
106. Egenburg IZ (1978) Russ. Chem. Rev. 43: 470

107. Hopf H (1982) In: Landor SR (ed) The chemistry of the allenes, vol 2, Academic, London, p 563
108. Skattebøl L (1967) Tetrahedron 23: 1107. The rearrangement is profoundly effected by substituents: Holm KH, Skattebøl L (1984) Acta Chem. Scand. B38: 783
109. Reinarz RB, Fonken GJ (1973) Tetrahedron Lett. 4591; cf. Minter DE, Fonken GJ, Cook FT (1979) Tetrahedron Lett., 711
110. Holm KH, Skattebøl L (1974) Tetrahedron Lett. 2347; cf. Skattebøl L (1984) Acta Chem. Scand. B38: 783; Brun R, Grace DSB, Holm KH, Skattebøl L (1986) Acta Chem. Scand., B40: 21
111. Baird MS, Jeffries I (1986) Tetrahedron Lett. 27: 2493
112. Fleischhauer I, Brinker UH (1986) Chem. Ber. 120; cf. Brinker UH, Ritzer J (1981) J. Am. Chem. Soc. 103: 2116
113. Baird MS, Reese CB (1976) Tetrahedron Lett. 2896
114. Brinker UH, Fleischhauer I (1979) Angew. Chem. 91: 424, (1979) Angew. Chem. Int. Ed. Engl. 18: 396
115. Fleischhauer I, Brinker UH (1983) Tetrahedron Lett. 24: 3205
116. Brinker UH, Fleischhauer I (1981) Tetrahedron 37: 4495
117. Brinker UH, Fleischhauer I (1980) Angew. Chem. 92: 314, (1980) Angew. Chem. Int. Ed. Engl. 19: 304; cf. Brinker UH, Fleischhauer I (1986) Chem. Ber. 119: 1244
118. Arct J, Skattebøl L (1982) Tetrahedron Lett. 23: 113
119. Brinker UH, Boxberger M (1983) J. Chem. Res. (S) 100
120. Baird MS (1971) J. Chem. Soc., Chem. Commun. 1145
121. Paquette LA, Zon G, Taylor RT (1974) J. Org. Chem. 39: 2676
122. Nilsen NO, Sydnes LK, Skattebøl L (1978) J. Chem. Soc., Chem. Commun. 128; cf. Nilsen NO, Skattebøl L, Sydnes LK (1982) Acta Chem. Scand. B36: 587
123. Baird MS, Kaura AC (1976) J. Chem. Soc. Chem. Commun. 356; cf. Boswell RF, Bass RG (1975) J. Org. Chem. 40: 2419
124. Arct J, Skattebøl L (1982) Acta Chem. Scand. B36: 593
125. Arct J, Skattebøl L, Stenstrom Y (1983) Acta Chem. Scand. B37: 681
126. Brinker UH, Haghani A, Gomann K (1985) Angew. Chem. 97: 235, (1985) Angew. Chem. Int. Ed. Engl. 24: 230
127. Holm KH, Lee DG, Skattebøl L (1978) Acta Chem. Scand. B32: 693
128. Sydnes LK, Skattebøl L (1978) Acta Chem. Scand. B32: 547; cf. Kitatani K, Hiyame T, Nozaki H (1977) Bull. Chem. Soc. Japan 50: 1600, 2158; Yamamoto H, Kitatani K, Hiyama T, Nozaki H (1977) J. Am. Chem. Soc. 99: 5816
129. Ransom CJ, Reese CB (1975) J. Chem. Soc., Chem. Commun. 970
130. Billups WE, Leavell KH, Chow MY, Lewis ES (1972) J. Am. Chem. Soc. 94: 1770
131. Billups WE, Baker BA, Chow WY, Leavell KH, Lewis ES (1975) J. Org. Chem. 40: 1702
132. Billups WE, Shields TC, Chow WY, Deno NC (1972) J. Org. Chem. 37: 3676
133. Surmina LS, Formanovskii AA, Bolesov IG (1978) Zh. Org. Khim. 14: 883; (1978) J. Org. Chem. USSR 14: 821
134. Lindsay DG, Reese CB (1965) Tetrahedron 21: 1673
135. Tarakanova AV, Grishin JK, Vashakidze AG, Milvitzkaya EM, Plate AF (1972) Zh. Org. Khim. 8: 1619
136. Banwell MG (1982) J. Chem. Soc., Chem. Commun. 847
137. Skattebøl L, Nilsen NO, Myhren F (1986) Acta Chem. Scand. B40: 782
138. Billups WE, Blakeney AJ, Chow WY (1971) J. Chem. Soc., Chem. Commun. 1461
139. Billups WE, Chow WY (1973) J. Am. Chem. Soc. 95: 4099
140. Billups WE, Casserly EW, Arney BE (1984) J. Am. Chem. Soc. 106: 440
141. Ippen J, Vogel E (1974) Angew. Chem. 86: 780, (1974) Angew. Chem. Int. Ed. Engl. 13: 736
142. Davalian D, Garratt PJ, Koller W, Mansuri MM (1980) J. Org. Chem. 45: 4183
143. Billups WE, Chamberlain WT, Asim MJ (1977) Tetrahedron Lett. 571
144. Halton B, Randall CJ (1982) Tetrahedron Lett. 23: 5591
145. Baird MS, Nethercott W (1983) Tetrahedron Lett. 24: 605; cf. Baird MS, Hussain HH, Nethercott W (1986) J. Chem. Soc., Perkin I, 1845
146. Baird MS, Buxton SR, Whitley JS (1984) Tetrahedron Lett. 25: 1509

147. Review: Baird MS (1988) Topics Curr. Chem. 144: 137 and references cited therein
148. Mißlitz U, de Meijere A (1980) In: Houben-Weyl, vol E19, Thieme, Stuttgart, p 664 and references cired therein
149. Slougui N, Rousseau G (1987) Tetrahedron Lett. 28: 1651
150. Tobey SW, West R (1968) J. Am. Chem. Soc. 88: 2478; see also Sepiol J, Soulen RL (1975) J. Org. Chem. 40: 3791; Glück C, Poignée V, Schwager H (1987) Synthesis 260
151. Law DF, Tobey SW (1968) J. Am. Chem. Soc. 90: 2376; Seitz G, von Gemmern R (1987) Synthesis 953; Neidlein R, Poignée V, Kramer W, Glück C (1986) Angew. Chem. 98: 735, (1986) Angew. Chem. Int. Ed. Engl. 25: 731
152. Reviews: Deem ML (1972) Synthesis 675; (1982) 802 and references cited therein. For a recent interesting reaction leading to indolizines see: Smith KA, Waterman KC, Streit-wieser A Jr (1985) J. Org. Chem. 50: 3360
153. Tobey SW, West R (1964) J. Am. Chem. Soc. 86: 4215; West R, Zecher DC, Tobey SW (1970) J. Am. Chem. Soc. 92: 168; for recent examples and applications see: Wadsworth D, Geer S, Detty M (1987) J. Org. Chem. 52: 3662; Eicher T, Huck V, Schneider V, Veith M (1989) Synthesis 367; Eicher T, Schneider U (1989) Synthesis 372
154. Musigmann K, Mayr H, de Meijere A (1987) Tetrahedron Lett. 28, 4517
155. For reviews see: de Meijere A (1984) Bull. Soc. Chim. Belges 93: 241; de Meijere A (1987) Chem. Britain 23: 865; de Meijere A (1987) In: Oglobin KA (ed) Modern problems of organic chemistry, No. 9, Leningrad University Press, Leningrad, p 5; de Meijere A, Wessjohann L (1990) Synlett (20)
156. Nefedov OM, Menchikov LG, Djachenko AI, Agre CA (1986) Z. Vses. Khim Obshestva 31: 182
157. Jonczyk A, Dabromski M, Wozniak W (1983) Tetrahedron Lett. 24: 1065
158. Djachenko AI, Agre CA, Rudashevskaja TJ, Shafran RN, Nefedov OM (1984) Izv. Akad. Nauk. SSR, Ser. Khim., 2820. For a review cf. Binger P, Büch HM (1987) Topics Curr. Chem. 135: 77 and references cited therein
159. Novokreshchennyh AI, Mochalov SS, Shabarov JS (1978) Zh. Organ. Khim. 14: 546, (1978) J. Org. Chem. USSR 14: 505
160. Henseling KO, Weyerstahl P (1975) Chem. Ber. 108: 2803
161. Varakin GS, Kostikov RR, Ogloblin KA (1983) Zh. Organ. Khim. 19: 1768
162. Tishchenko IT, Kulimkovitch OG, Glasov JV (1975) Zh. Organ. Khim. 11: 581, (1975) J. Org. Chem. USSR 11: 579
163. Kobayashi Y, Taguchi T, Morikowa T, Takase T, Takanashi H (1962) J. Org. Chem. 47: 3232
164. Woodward RB, Hoffmann R (1970) The conservation of orbital symmetry, Verlag Chemie, Weinheim
165. Parham WE, Yong KS (1968) J. Org. Chem. 33: 3947
166. Parham WE, Yong KS (1970) J. Org. Chem. 35: 683
167. Bessiere Y, Schlosser M (1976) Helv. Chim. Acta 59: 969
168. Schlosser M (1978) Tetrahedron 34: 3
169. Warner P, Palmer R (1980) Tetrahedron Lett. 21: 145
170. Reese CB, Shaw A (1970) J. Chem. Soc. D 1365
171. Loozen HJ, Robbin WM, Richter TL, Buck HM (1976) J. Org. Chem. 41: 384
172. Reese CB, Stebles MRD (1972) Tetrahedron Lett. 4427
173. Arct J, Prawda A, Kozyriev V (1978) Bull. Acad. Pol. Sci. 24: 523
174. Ito S, Ziffer H, Bax A (1986) J. Org. Chem. 51: 1130
175. Christl M, Freitag G (1976) Angew. Chem. 88: 508, (1976) Angew. Chem. Int. Ed. Engl. 15: 493; cf. Christl M, Freitag G, Brüntrup G (1978) Chem. Ber. 111: 2320
176. Danheiser RL, Morin JM, Basak MYA (1981) Tetrahedron Lett. 22: 4205
177. Reese CB, Shaw A (1970) J. Chem. Soc., Chem. Commun. 1365
178. Dhanak D, Kuroda R, Reese CB (1987) Tetrahedron Lett., 28: 1827
179. Skattebøl L, Boulette B (1966) J. Org. Chem. 31: 81
180. Skattebøl L (1966) J. Org. Chem. 31: 1554
181. Skattebøl L (1970) J. Org. Chem. 35: 3200
182. Baird MS, Hussein HH (1988) J. Chem. Res. Synop. 292

183. Groves JT, Ma RW (1974) Tetrahedron Lett. 909
184. Ishihara T, Kudaka T, Ando T (1984) Tetrahedron Lett. 25: 4765
185. Neureiter NP (1959) J. Org. Chem. 24: 2044
186. Varakin GS, Kostikov RR (1986) Sovremennye problemy organicheskoi khimii, No. 8, Leningrade State University, p 76; cf. Wong HNC, Hon M-Y, Tse C-W, Yip Y-C, Tanko J, Hudlicky T (1989) Chem. Rev. 89: 165
187. Fields R, Haszeldine RN, Peter D (1969) J. Chem. Soc. C 165
188. Duffey DC, Minyard JP, Lane RH (1966) J. Org. Chem. 31: 3865
189. Clifford RP, Holbrook KA (1972) J. Chem. Soc., Perkin II 1972
190. Ferrero JC, Cosa JJ, Staricco EH (1972) J. Chem. Soc., Perkin II 2382
191. Ioffe AI, Nefedov OM (1974) Izv. Akad. Nauk SSR, Ser. Khim., 1536; (1974) Bull. Acad. Science USSR, 23: 1455
192. Baird MS, Lindsay DG, Reese CB (1969) J. Chem. Soc. C 1173
193. Ando T, Hosaka H, Yamanaka H, Funasaka W (1969) Bull. Chem. Soc. Japan 42: 2013
194. Baird MS (1976) J. Chem. Soc., Perkin I 54
195. Windberg HE (1959) J. Org. Chem. 24: 264
196. Nefedov OM, Agaveljan AS (1972) Izv. Akad. Nauk. SSR, Ser. Khim 481; cf. Demhlow EV (1978) Tetrahedron 28: 175
197. Reese CB, Shaw A (1972) J. Chem. Soc., Chem. Commun. 271
198a. Jefford CW, Mareda J, Gehret J-CE, Kabengele T, Graham WD, Burger U (1976) J. Am. Chem. Soc. 98: 2585; cf. Katz TJ, Cerefice S (1969) J. Am. Chem. Soc. 91: 2405; 198b. Jefford CW, Kabengele T, Kovacs J, Burger U (1974) Tetrahedron Lett. 257; 198c. de Selms RS, Combs CM (1972) J. Chem. Soc., Chem. Commun. 271
199. Christl M, Freitag G (1976) Angew. Chem. 88: 508, (1976) Angew. Chem. Int. Ed. Engl. 15: 493; cf. Christl M, Herzog C, Brückner D, Lang R (1986) Chem. Ber. 119: 141
200. Kostikov RR, Varakin GS, Ogloblin KA (1983) Zh. Organ. Khim., 19: 1625, (1983) J. Org. Chem. USSR 19: 1438; cf. Magarian HA, Milton S, Natarelly G (1972) J. Pharmaceutol. Sci. 61: 1216; Dehmlow EV, Schönefeld J (1971) Liebigs Ann. Chem. 744: (2
201. McElvain SM, Weyna PL (1959) J. Am. Chem. Soc. 81: 2579
202. Asahara T, Ono K, Tanaka T (1971) Bull. Chem. Soc. Japan 44: 1130
203. Schlosser M, Spahic B, Tarchini C, Le Van Chan (1975) Angew. Chem. 87: 346, (1975) Angew. Chem. Int. Ed. Engl. 14: 365; cf. Spahic V, Thimy Thy T, Schlosser M (1980) Helv. Chimica Acta 63: 1236
204. Schlosser M, Spahic B (1980) Helv. Chimica Acta 63: 1223
205. Horspool WM, Sutherland RG, Thomson BJ (1971) J. Chem. Soc. C 1558
206. Shimizu N, Nishida S (1977) Chem. Lett. 839
207. Kostikov RR, Molchanov AP (1978) Zh. Organ. Khim. 14: 1108, (1978) J. Org. Chem. USSR 14: 1032
208. Molchanov AP, Kostikov RR (1984) Zh. Organ. Khim. 20: 2118, (1984) J. Org. Chem. USSR 20: 1930
209. For a recent review see Hudlicky T, Kutchan TM, Naqvi SM (1985) Org. Reactions 33: 247; cf. Paquette LA, Doherty AM (1987) Polyquinane chemistry, Springer-Verlag, Heidelberg
210. Ketley AD, Berlin AJ, Gorman E, Fisher JP (1966) J. Org. 31: 305
211. Dolbier WR, Sellers SF (1982) J. Am. Chem. Soc. 104: 2494. A review of the authors works on thermal rearrangements of gem-difluorocyclopropanes covering, inter alia, cyclopropane thermolysis, methylene cyclopropane and spiropentane rearrangements, vinylcyclopropane and cyclopropylcarbinyl isomerizations has been published: Dolbier WR (1981) Acc. Chem. Res. 14: 195
212. Dolbier WR, Sellers SF (1982) J. Org. Chem. 47: 1
213. Roth WR, König J, Stein K (1970) Chem. Ber. 103: 426
214. Jenneskens LW, de Wolf WH, Bickelhaupt F (1986) Chem. Ber. 119: 754
215. Dehnlow EV, Balschukat D, Schmidt PP, Krause R (1986) J. Chem. Soc., Chem. Commun. 1435
216. The extensive literature on this process has been summarized by Gajewski JJ (1981) Hydrocarbon thermal isomerizations, Academic, New York, p 90

217. Slafer WD, English AD, Harris DD, Shellhamer DF, Meshishnek MS, Aue DH (1975) J. Am. Chem. Soc. 97: 6638
218. Chesick JP (1963) J. Am. Chem. Soc. 85: 2720
219. Crawford RJ, Tokunaga H (1974) Canad. J. Chem. 52: 4033
220. Kirmse W, Murawski H-R (1977) J. Chem. Soc., Chem. Commun. 122
221. Dolbier WR, Fielder TH (1978) J. Am. Chem. Soc. 100: 5577
222. Gajewski JJ (1971) J. Am. Chem. Soc. 93: 4450
223. Kostikov RR, Molchanov AP (1979) Doklady Akad. Nauk USSR 246: 1377
224. Dolbier WR, Seabury M, Daly D, Smart BE (1986) J. Org. Chem., 51: 974
225. Höhn J, Pickardt J, Weyerstahl P (1983) Chem. Ber. 116: 798
226. Weyerstahl P, Blume G (1972) Tetrahedron 28: 5281
227. Makosza M, Gajos I (1974) Roczniki Chem. 48: 1883
228. de Angelis F, Gambacorte A, Nicoletti R, (1976) Synthesis 798; Dehmlow EV, Franke K (1979) Liebigs Ann. Chem. 1456
229. MacDonald TL (1978) Tetrahedron Lett. 4201
230. Hirao T, Harano Y, Yamana Y, Ohshiro Y, Agawa T (1983) Tetrahedron Lett. 24: 1255
231. Hirao T, Fuijhara Y, Kurokawa K, Oshiro Y, Agawa T (1986) J. Org. Chem. 51: 2830
232. Janovskaja LA, Dambrovskij VA, Khusid AK (1980) Cyclopropanes with functional groups, Nauka, Moscow, p 221
233. Hashem MA, Hülskämper R, Weyerstahl P (1985) Chem. Ber. 118: 840; Weyerstahl P, Hashem MA (1987) Chem. Ber. 120: 449; cf. Sasaki T, Kamematsu K, Yukimoto Y (1974) J. Org. Chem. 39: 155
234. Christl M, Herzog C, Brückner D, Lang R (1986) Chem. Ber., 119: 141
235. Christl M, Freitag G, Brüntrup G (1978) Chem. Ber. 111: 2320
236. Christl M, Lechner M (1982) Chem. Ber. 115: 1
237. Christl M, Nusser R, Herzog C (1988) Chem. Ber. 121: 309
238. Christl M, Lang R, Herzog C (1986) Tetrahedron 42: 1585

Through-Bond Modulation of Reaction Centers by Remote Substituents

Tse-Lok Ho

The NutraSweet Co., 601E, Kensington Rd., Mt. Prospect, IL 60056, USA

Table of Contents

A reaction center is under influence not only of substituent(s) which it bears, but also by those at a more distant location. The influence is reflected in its reactivities towards different reagents. This article examines the structural-reactivity relationships in terms of the number of intervening bonds and the nature of both the reaction center and the substituent(s). The polarity alternation rule is the basis for this reactivity assessment. Through-space interactions are not discussed.

Topics in Current Chemistry, Vol. 155
© Springer-Verlag Berlin Heidelberg 1990

1 Introduction

In organic chemistry it is well known that a substituent directly linked to a reaction center has a profound effect on the latter's reactivity. The influence of a remote substituent has also received much attention as more organic molecules are being scrutinized. Thus, neighboring group participations in solvolysis are generally observed. Certainly, the full attribution of a substituent must be the sum of electronic, steric, and stereoelectronic factors.

This article discusses many organic reactions as affected electronically by substituents situated more than one bond away. More specifically, the polarity alternation rule forms the basis for evaluation of organic reactivity [1]. The concept, with its simplicity and broad applicability, deserves to be the prime and initial focus for a student to learn about organic reactivity.

The polarity alternation rule (PAR) considers two kinds of substituents. The donors are those having unshared electronic pairs or π-electrons, and $+I$ groups. These include OH, OR, OCOR, NH_2, NRR', N(R)COR', SH, SR, halogens and alkyl groups. The donor properties of the alkyl groups may reflect the existence of hyperconjugation. On the other hand, the acceptors are electron sinks, i.e. polarizable π-bonds, atoms with empty orbitals, and $-I$ groups. Examples of acceptors are C=O (aldehydes, ketones, carboxylic acid derivatives), CN, SO_2, NO_2, SiR_3.

The coulombic forces determine linkage of a donor to an acceptor, and vice versa. In other words, direct covalent bonding of a donor or acceptor to another atom of its kind is unfavorable.

Since hydrocarbon subunits (methyl, methylene and methine groups) are not polarized to a great extent, their nature can be defined by a polar substituent. The high acidity of the α-hydrogen atoms of carbonyl compounds, nitriles, sulfones, and nitroalkanes follows from polarity alternation, the carbon atoms being a donor next to the acceptor substituent.

Theoretical appraisal of polar effects [2] has indicated a charge alternation associated with back-donation of lone pair electrons of donors. Coulombic interactions seem to be sufficient to stabilize certain acyclic systems and therefore it is not necessary to invoke resonance effects [3].

Long range electronic effects sometimes can be discerned. Thus, the striking difference in basicity [4] of 4-cyclopropylpyridine [pKa 6.44] and 4-(2-carbethoxy cyclopropyl)pyridine [pKa 5.58] is attributable to the field effect and through-bond interaction.

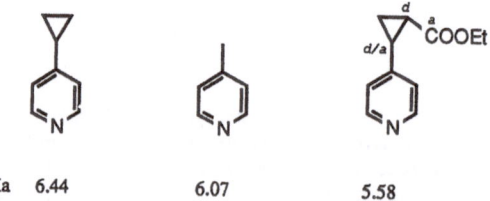

pKa   6.44   6.07   5.58

Lithiation study of xylenes indicates a much higher reactivity of the *meta* isomer (m- $\gg$ o- $>p$-). On the other hand, only the *meta* isomer of methylacetophenone resists metalation at the sidechain. These observations are totally consistent with the polarity alternation rule which can be used to designate donor sites [5].

Another example concerns with the release of calcium ion from its chelate with phenylglycine analogs which contain an *o*-nitrobenzhydrol system as triggered by illumination [6]. The photochemical transformation of the latter moiety into a nitrosobenzophenone forces the amino nitrogen to adopt a different conformation in which its acetic acid chains can no longer bind the metal ion cooperatively. In the benzophenone the amino and the ketone groups are PA-related (conjugation).

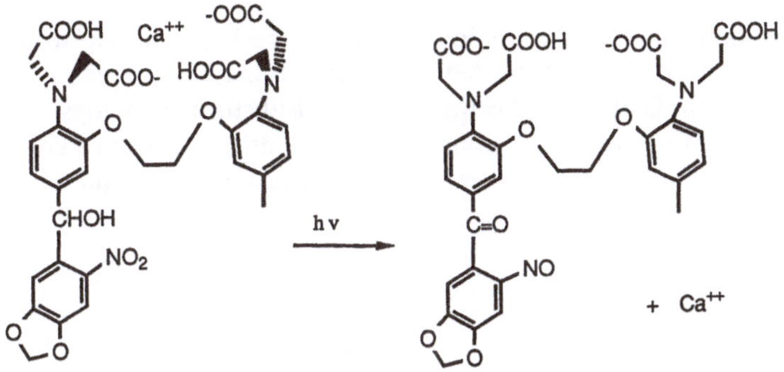

(1)

Interesting situations develop when multiple (>2) substituents are placed in fairly close proximity. Depending on the donor or acceptor nature of the substituents and that of the carbon center in question, and the number of bonds separating them, there will be an activating or deactivating effect. Typical scenarios are summarized in the following diagrams.

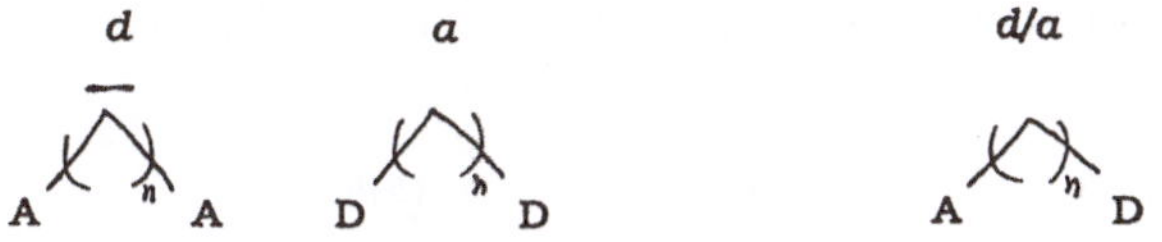

d	*a*	*d/a*
n = odd	favorable (conjoint)	unfavorable
n = even	unfavorable (disjoint)	favorable

In the favorable cases, polarity alternation prevails along the atomic chains within which the reaction centers reside. On the contrary, unfavorable interactions between the substituents and the reaction centers are due to disruption of the polarity alternation. As will be seen later, the favorableness/unfavorableness is in the thermodynamic context. Recent calculations [7] of energy changes for hydride transfer and chloride transfer to $^+CH_2OH$ from CH_3SH and $ClCH_2SH$, respectively, manifest a stabilizing interaction of the geminal Cl, OH relative to the Cl, SH pair.

It is possible, or even desirable, to activate a chemical species to undergo reactions by setting up an unfavorable situation. In short, *reaction activation can be achieved by either polarity alternation accentuation or polarity alternation disruption.*

The PAR is easy to apply, and the crucial need is only to identify correctly the donor or acceptor nature of a polar substituent. It should be mentioned that an interesting phenomenon called contrapolarization [8] can often be observed. This phenomenon pertains to a role change of an atomic center during a reaction step, it occurs ubiquitously in redox processes. Contrapolarization is unintentional and transitory umpolung.

2 Substitution Reactions

2.1 Aromatic Substitutions

The most familiar set of organic reactions is perhaps the electrophilic aromatic substitutions. For monosubstituted benzenes the major products from the process are either *o*- or *p*-disubstituted benzenes or *m*-disubstituted analogs.

The *o,p*-directing groups are donors and the *m*-directing groups are acceptors. The cause-effect is readily revealed by the polarity alternating sequences shown in the following formulas.

The "unusual" *meta*-bromination of phenols and their methyl ethers [9] under superacid conditions is quite expected because *O*-protonation inverts polarity of all the nuclear carbon atoms. This reaction is reminiscent of the better known electrophilic substitutions of aniline in acidic media.

It has been shown that *t*-butylation of acetanilide [10] gives the 1,4-isomer as the kinetic product which is then converted into the 1,3-isomer. The isomerization is a 1,2-alkyl migration within the C-protonated intermediate which has a perfect set-up of donor and acceptor sites, and the two substituents of 3-(*t*-butyl)acetanilide are conjoint in polarity.

$$(2)$$

A significant Friedel-Crafts acylation-dealkylation sequence has been observed [11]. The powerful directing effect of an alkoxy group excludes the pathway leading to an intermediate with a condensed ring system.

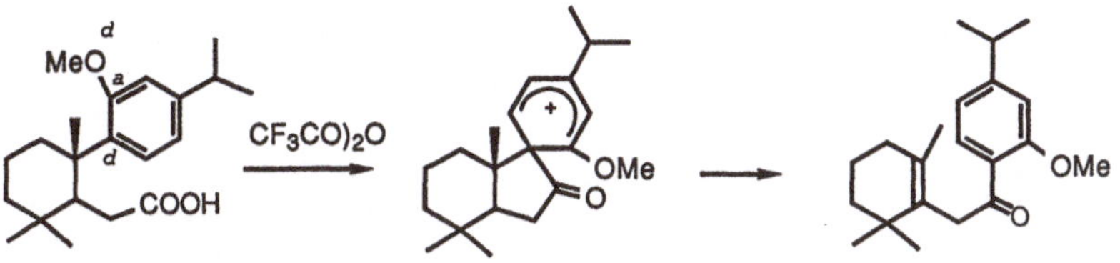

(3)

According to polarity alternation the acceptor substituted benzenes should also undergo *ipso* substitutions [12] Indeed such reactions occur when A = SiMe$_3$, SnR$_3$, HgR. A useful application of the silyl group as site determinant is found in a synthesis of 8-methoxy-*N*-methyl-1,2,3,4-tetrahydroisoquinoline [13]. The donor character at the *ortho* position of the methoxy function is accentuated and steric disfavor is overcome.

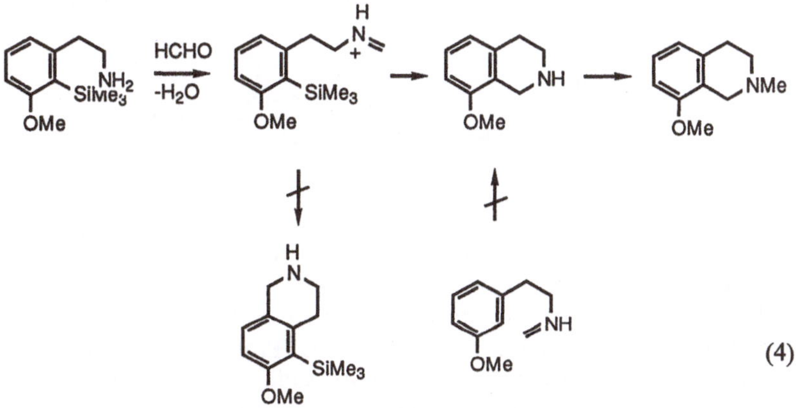

(4)

Intramolecular alkylation of phenol with diazoketone [14] can be analyzed as the following: The carbenoid center acts first as an acceptor and then a donor. Since the resulting cyclopropane derivative is vicinally substituted with an acceptor (C=O) and a donor (enol), fragmentation follows instantaneously (*vide infra*).

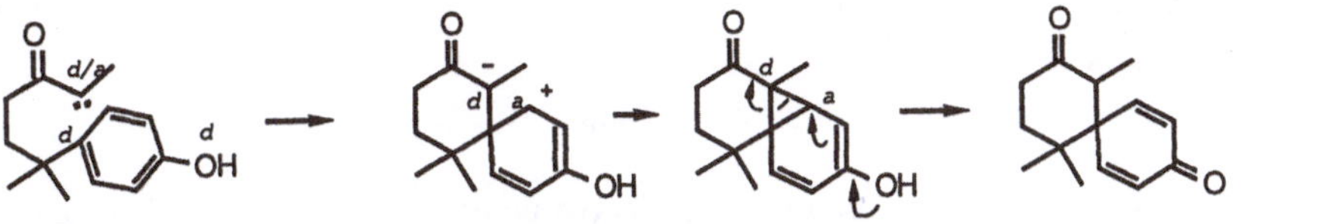

(5)

An indirect 3-alkylation of pyridine [15] is based on accentuation of the donor character of C-3 by reductive dearomatization.

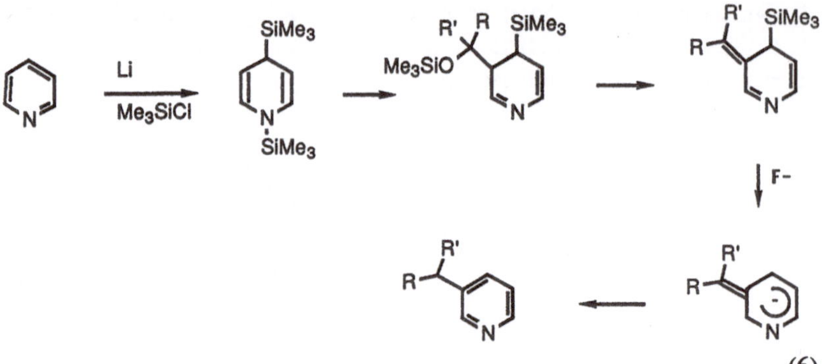

(6)

Nucleophilic substitutions of simple aromatic compounds which formally involve a hydride displacement are difficult to achieve because of the poor leaving group and the high electron density of the aromatic nucleus which repels approach of a nucleophile. However, π-electron deficient aromatic compounds such as metal carbonyl complexes are susceptible to attack by certain carbon nucleophiles. Studies of this chemistry have shown [16] an opposite regioselectivity to the corresponding electrophilic substitutions, in agreement with the polarity alternation rule.

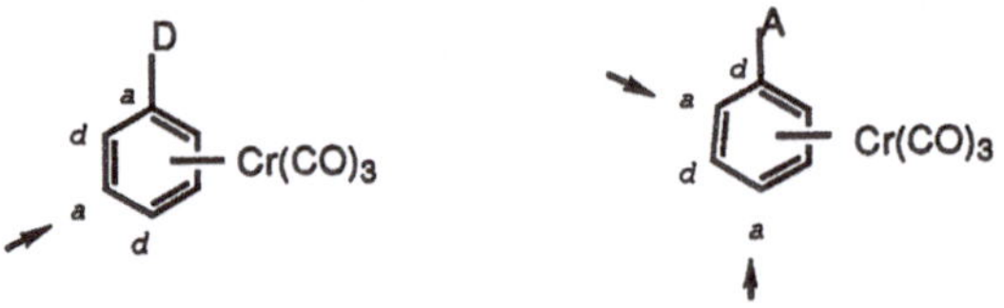

(D)

The possibility of readily introducing a *meta* sidechain to an anisole has enabled design an unusual synthesis of acorenone [17].

The σ-adducts from such reactions are generally subjected to oxidative decomplexation to generate the aromatic products. Protonation/deuteration of an adduct derived from a donor-substituted benzene could result in a net *meta* displacement of the donor [18]. *Ipso* substitution has also been revealed when the σ-complex is warmed.

Reaction of η^6-arene-η^5-cyclopentadienyliron cations with ketone enolates results in regiospecific attack at the *ortho* position of the acceptor substituent [19].

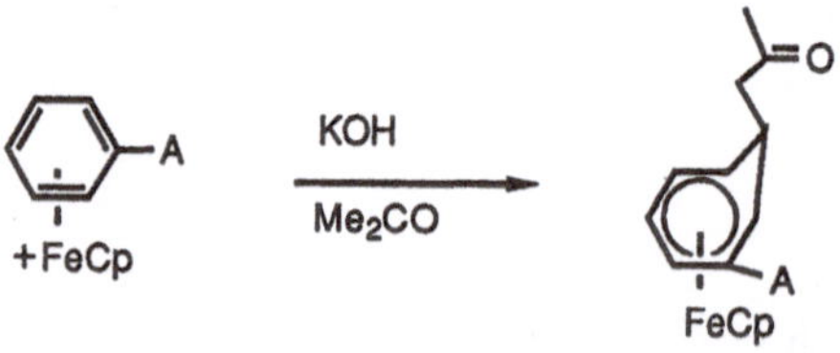

(7)

Relevant to the unveiling of the intrinsic electronic effects of various substituents is the selective lithiation of the fluoroanisole-Cr(CO)$_3$ complexes [20]. The chelation control by the etheral oxygen atom which prevails in uncomplexes aromatic ethers becomes far less important.

$$\text{(structure)} \xrightarrow[\text{RX}]{\text{nBuLi;}} \text{(structure)} \tag{8}$$

Nitration of dibenzofuran at C-3 as opposed to other electrophilic substitutions such as acetylation at C-2 has been attributed to the intervention of a charge-transfer process [21]. The C-N bond formation step is mechanistically closer to the nucleophilic process, the aromatic moiety being the electron-deficient species. It is understandable that $\cdot NO_2$. attacks at a nuclear carbon which is *meta* to the oxygen donor.

$$\text{(structure)} \rightleftharpoons \text{(structure)} \tag{9}$$

The vicarious substitutions of nitroarenes [22] involve nucleophiles which contain a leaving group X to enable elimination of HX from the σ-adducts. As expected, the substitution is *o-/p-*to the nitro acceptor.

$$\text{(structure)} \xrightarrow{XYCH^-} \text{(structure)} \xrightarrow{-HX} \text{(structure)} \xrightarrow{H^+} \text{(structure)} \tag{10}$$

An interesting aspect of these substitutions is that the HX-elimination demands a contrapolarization at the nuclear carbon that acts as an acceptor during the addition.

As expected, N-pentafluorophenylpiperidine N-oxide undergoes substitution at positions *ortho* or *para* to the nitrogen atom [23]. Secondary amines give the *ortho* products because of their capability for hydrogen bonding in the transition state.

Remarkable regioselectivity is exhibited in aromatic substitutions via aryne intermediates. When there is a donor group at the meta position, apparent *ipso* substitution occurs [24]. The polarity alternation sequence indicates an electronic cause.

$$\text{(structure)} \xrightarrow{NH_2^-} \text{(structure)} \xrightarrow{RNH_2} \text{(structure)} \tag{11}$$

D= OMe, NMe$_2$, Cl

Interestingly, in the reaction of 2-(3-chlorophenyl)-4,4-dimethyloxazoline with an alkyllithium [25], the deprotonation is chelation-controlled, but the addition of R$^-$ follows PAR. A reverse regioselectivity for alkylcuprate addition may well be explained by a complex-induced proximity effect which has been invoked to rationalize the alkyllithium additions. It is known that copper coordinates exceptionally well with amino and imino ligands.

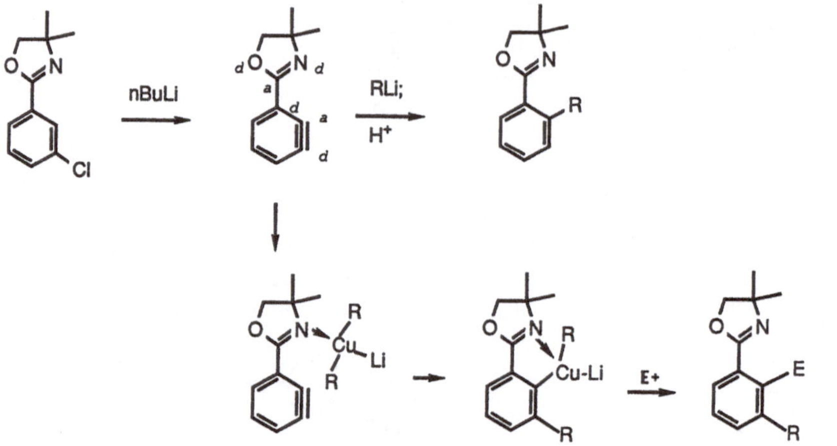

$$(12)$$

Cross-coupling of aryl halides exhibits fair to excellent selectivities [26]. Each of the two aryl halides contains a donor or acceptor group at the *para* position, respectively. The coupling sites are activated and the 4,4'-biphenyl products are thermodynamically most stable in electronic terms.

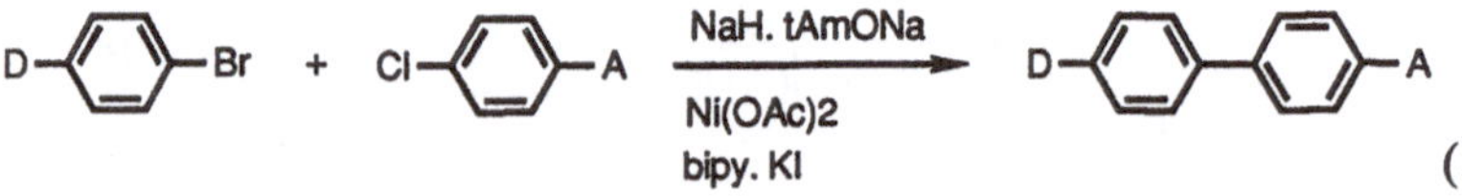

$$(13)$$

1,2-Dimethyl-4-nitroimidazole contains a better conjoint circuit than the 5-nitro isomer. Consequently it is not surprising that the latter compound undergoes isomerization to the former on exposure to iodomethane [27].

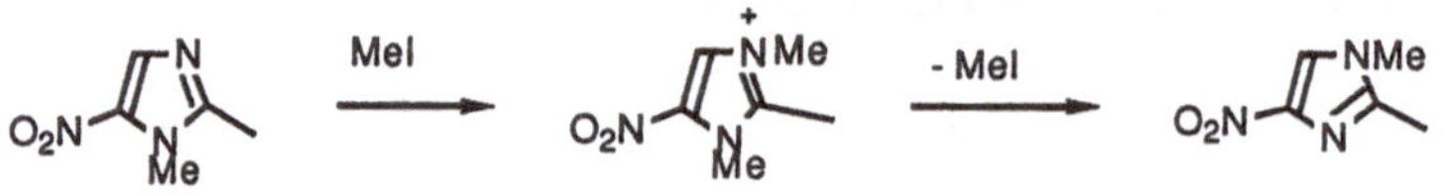

$$(14)$$

2.2 Aliphatic Substitutions

2,2,2-Trifluoromethyl iodide is a poor substrate for SN2 reactions [28]. The increased "donativity" of the methylene carbon as rendered by the fluorine atoms is reflected in its reluctance to enter a bonding relationship with a nucleophile. Similar reactivity patterns are known for chloromethyl phenyl sulfone [28] and chloromethyltrimethyl-silane. In these latter compounds the "reactive" center is directly linked to an acceptor group.

Benzylic halides and sulfonates show a wide range of reactivity towards nucleophiles. Activation and deactivation by *o-/p*-donors (e.g. OR) and acceptors (e.g. NO₂), respectively, are consistent with PAR. In each case the benzylic carbon atom is identified as acceptor or donor. The trends are also reflected in the relative acidities of the corresponding toluene derivatives.

In pyridine derivatives, α- and γ-picolines readily lose a proton from the methyl group. It is easy to find parallel reactivity in the picolines and the nitrotoluenes $\{N=CHNO_2\}$. Further development of the reactivity patterns includes a methanolytic study of several trichloromethylpyridines [29].

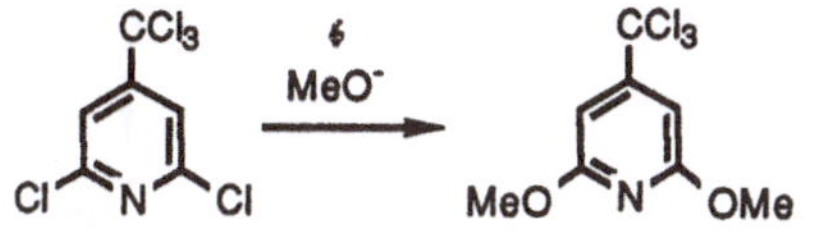

$$\text{(15)}$$

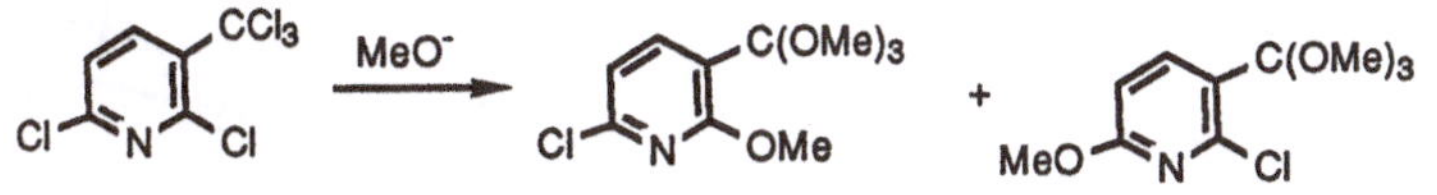

$$\text{(16)}$$

Vilsmeier reaction of 1,4-dihydrobenzoyl chlorides leads to *ipso*- and *meta*-substitution products. Aromatic aldehydes are isolated as a result of aerial oxidation [30]. The allylic acceptor (COX) is responsible for the double bond activation as well as the observed regioselectivity.

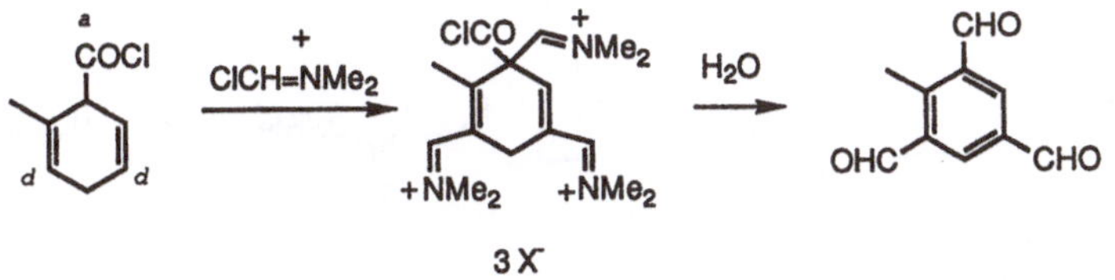

$$\text{(17)}$$

It may be of interest to point out that captodative [31] substituents on a carbon atom disfavor the removal of a proton attaching to it, reactions such as alkylation would occur preferentially at an alternative site by default. Thus a synthesis of yohimbine [32] via a Dieckmann cyclization for *E*-ring formation was facilitated by incorporating an extra methoxy group into one of the ester pendants.

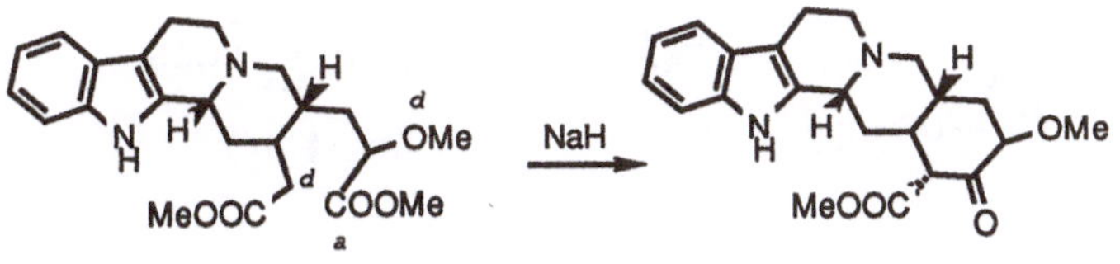

$$\text{(18)}$$

Another example of such effect is the increased regioselectivity for alkylation of certain bicyclic piperazinediones [33]. The stability difference of the two bridgehead carbanions is due to the third substituent (O vs. CH_2) {see Eq. 19}.

The dominance of an alkoxy group over a silyl group is manifested in the generation of amides from reaction of α-silyl enol ethers with chlorosulfonyl isocyanate, followed by hydrolysis. Simple vinylsilanes and alkynylsilanes undergo carbamidation [34].

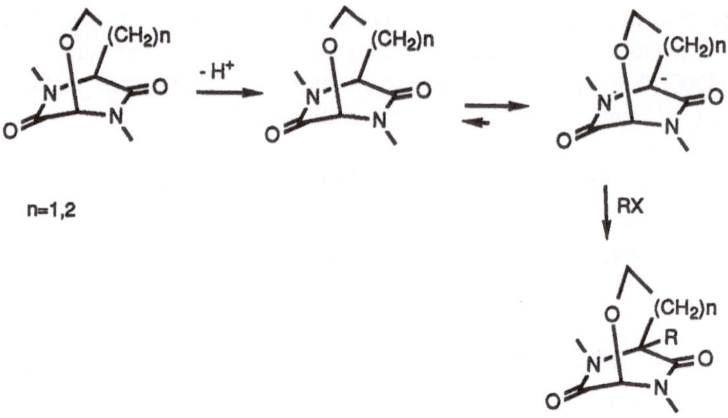

$$n=1,2 \qquad (19)$$

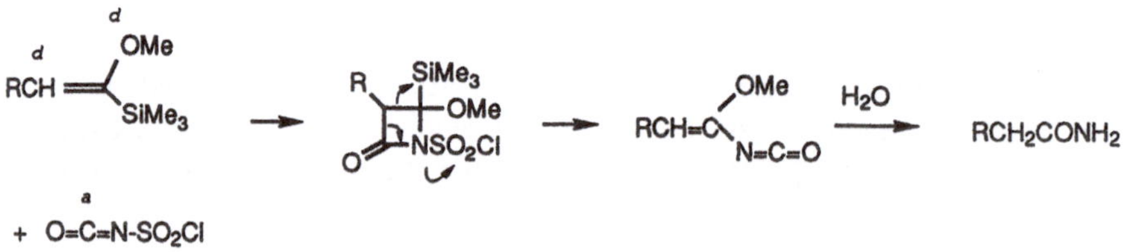

$$+ \; O=C=N-SO_2Cl \qquad (20)$$

$$+ \; O=C=N-SO_2Cl \qquad (21)$$

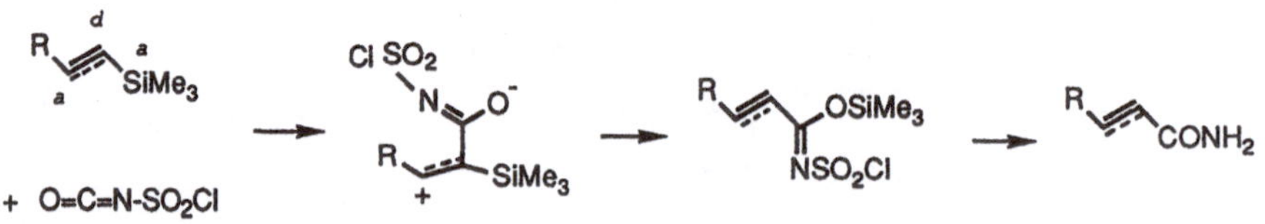

Substitutions at saturated carbon atoms that are subject to control by remote functionalities may be best illustrated by the ring opening of aziridines [35] and epoxides [36, 37].

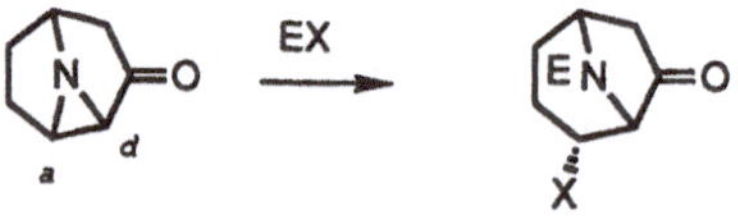

$$(22)$$

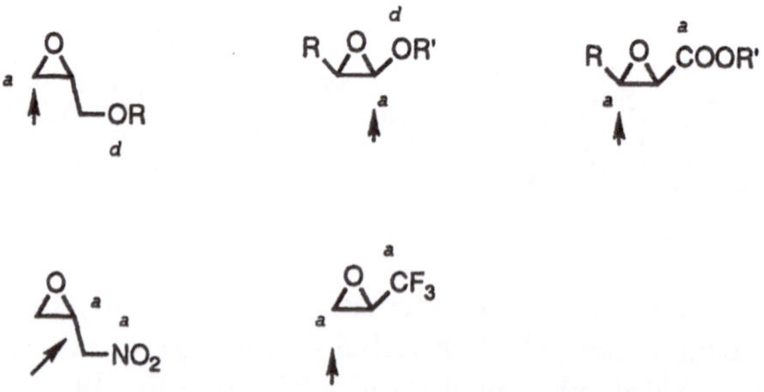

Notwithstanding the possible divergent reaction mechanisms for the two types of epoxides {direct displacement vs elimination-Michael addition sequence}, the polarity alternation rule reveals the correct regiochemistry. The same can be said for analysis of aromatic substitutions based on PAR or resonance considerations.

vic-Hydroxy azides and hence the corresponding hydroxy amines in which the hydroxy group is allylic may be prepared from epoxy azides [38].

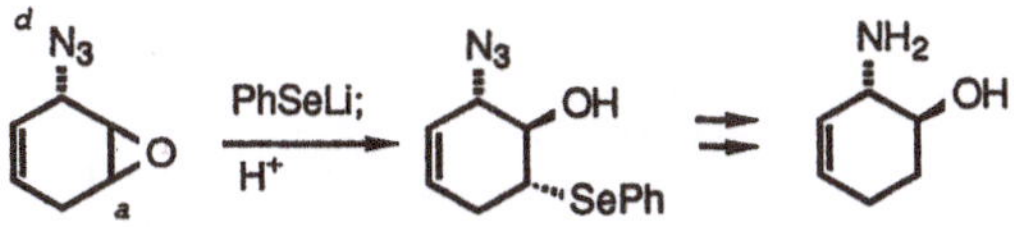

$$\text{(23)}$$

Another pertinent case is the formation of a tertiary fluoride from treatment of β,β-disubstituted α,β-epoxy nitriles [39].

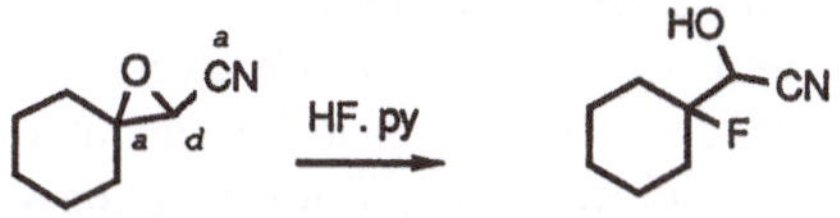

$$\text{(24)}$$

Regioselective alkylative opening of α,β-epoxy ketone oxime [40] and tosylhydrazones [41] with organocopper reagents is initiated by the generation of unsaturated azo compounds which undergo conjugate addition.

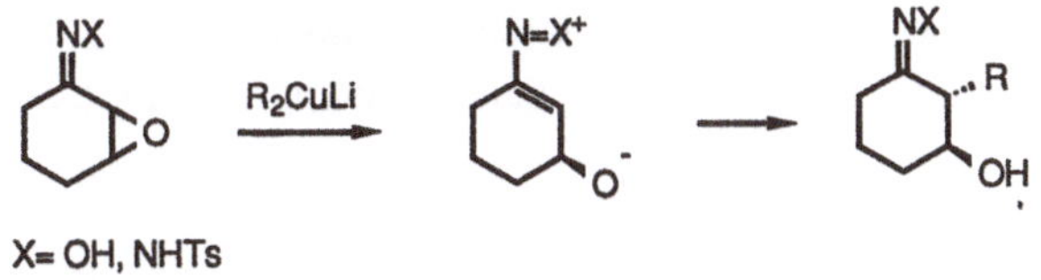

$$\text{(25)}$$

X= OH, NHTs

The reaction of epoxysilanes with organocuprate reagents has been employed in a synthesis of the silkworm pheromone bombykol [42]. The silyl group must be responsible for the interesting regioselectivity of the ring cleavage. The electrophilic center as shown is captodative (flanked by a donor oxygen and an acceptor silicon) and its destruction is thermodynamically driven.

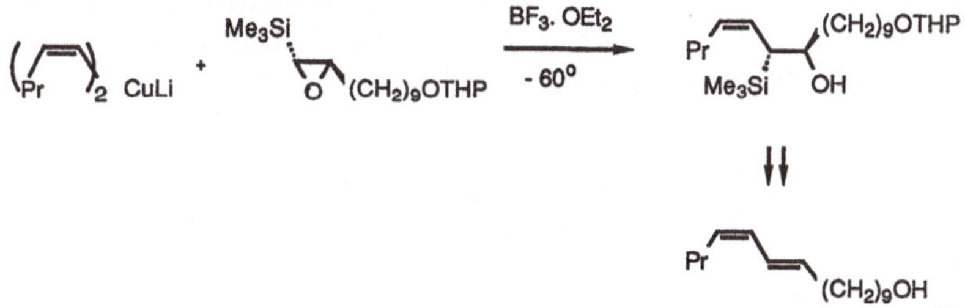

$$\text{(26)}$$

In an approach to the prostaglandins [43], seco-solvolysis of an epoxycyclopropane serves to establish the oxy functionality in the cyclopentane subunit and an E double bond in the side-chain. Opening of the epoxide that triggers the sequence of events is directed by the cyclopropyl group (donor) and the cyano group (acceptor).

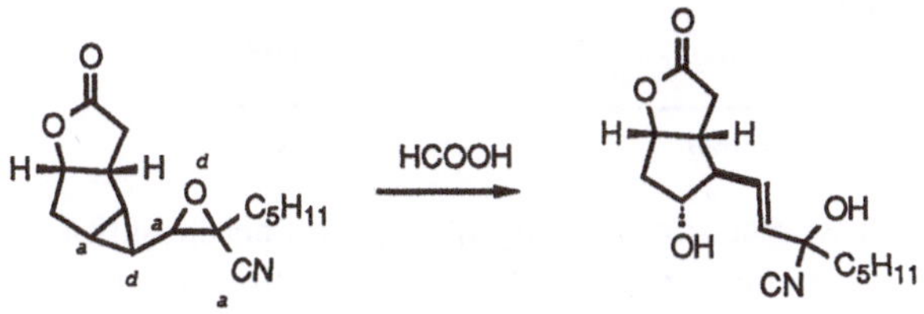

$$\text{(27)}$$

91

While the Baldwin rules [44] indicate both 5-*exo*-tri and 6-*endo*-tri modes of cyclization are favorable, only the latter is observed in the following reaction [45].

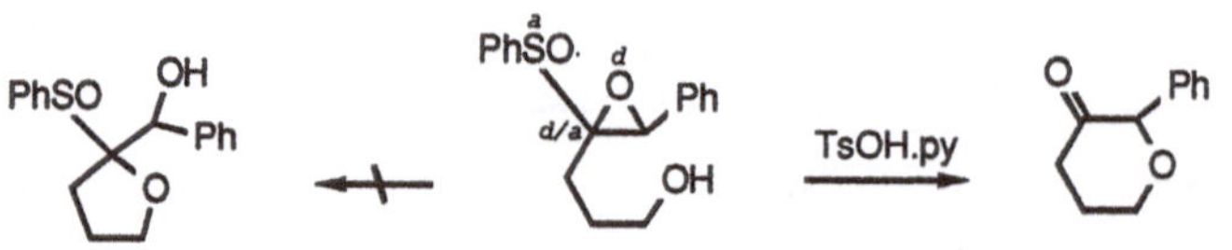

$$(28)$$

The cause for this regioselectivity may be the influence of the sulfoxide (acceptor) which renders its proximal carbon in the epoxide ring more donor like. Furthermore, ring opening at this site would not improve the unfavorable electronic status of the molecule, while attack on the distal carbon atom gives an intermediate in which the unfavorable captodative situation is dissolved on expulsion of phenylsulfenic acid, upon contropolarization at the sulfur atom.

Intramolecular opening of 2-azabicyclo[3.3.0]oct-7-ene-3 *endo*-carboxamide epoxide displays a dependence on the substituent at the nuclear nitrogen [46]. The conformational preference for a proximal attack by the amide function is counterbalanced by an accentuated polarity alternating sequence which enforces the acceptor role of the distal carbon of the epoxide ring when the nuclear nitrogen is part of a carbamate group.

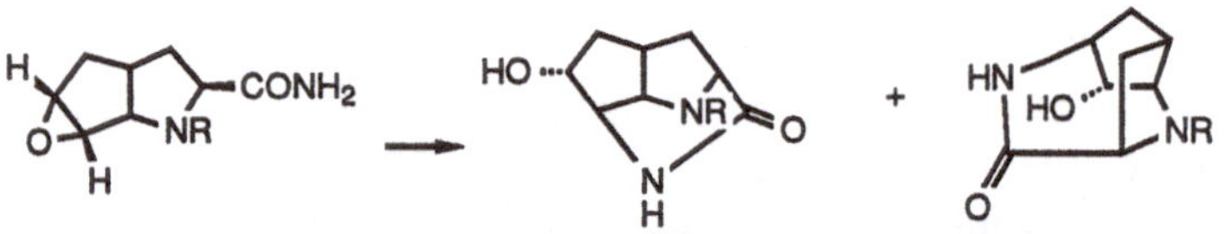

$$(29)$$

vic-Diol cyclic sulfates should prove more versatile than epoxides in synthesis as they contain two electrophilic centers [47]. The first reaction of a cyclic sulfate with a donor reagent is facile as the *a-a* arrangement is destroyed. It must be emphasized that these cyclic disulfates are more than synthetic equivalents of epoxides, the latter being unable to undergo cyclopropanation in one step.

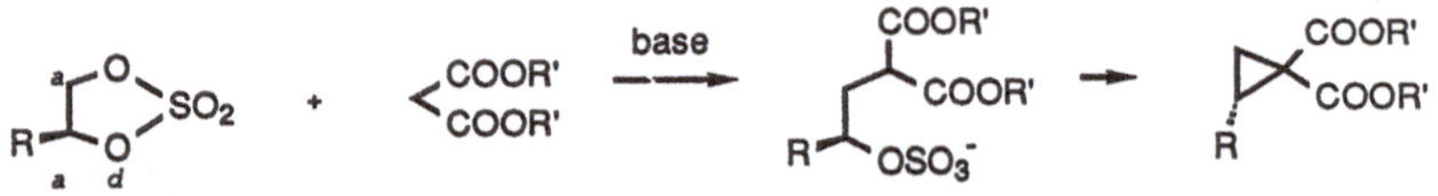

$$(30)$$

Regioselective enolization toward the nitrogen-bearing carbon in an oxazolinylmethyl ketone derived from aspartic acid has been observed [48]. The nitrogen atom of the heterocycle is conjoint with the carbonyl, therefore the creation of a conjugate system via enolization is favorable.

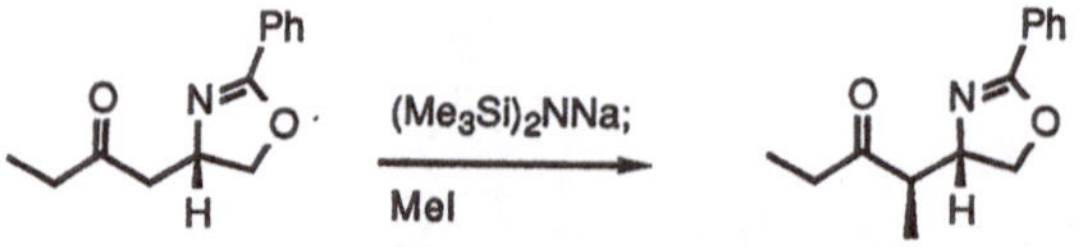

$$(31)$$

As implied in the above reaction, polarity alternation accentuation represents one strategy for activating organic molecules towards substitution. For example, there is a rate enhancement in the order of 10^5 in the ethylation of triphenylphosphine when ethyl iodide is first treated with a rhenium complex [49].

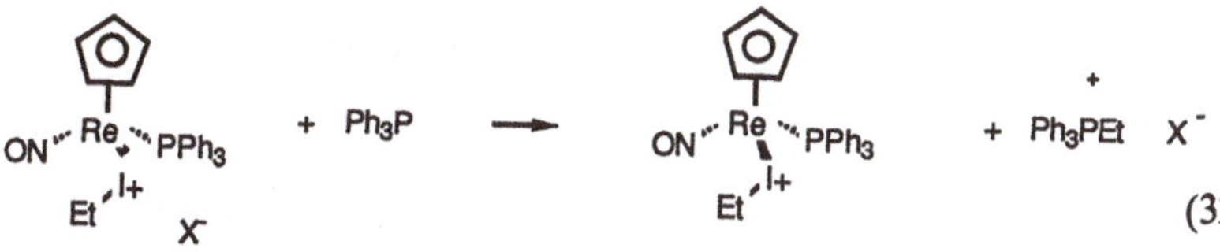

$$ \text{(32)} $$

Bistriflyl amines constitute a rare class of amino derivatives which undergo S_N2 reactions at the α-carbon [50]. The presence of two strong acceptor groups in the β position of this carbon greatly facilitate the C-N bond cleavage.

$$ PhCH_2NTf_2 + X^- \rightarrow PhCH_2X + Tf_2N^- \tag{33} $$

Ester hydrolysis, alcoholysis, and aminolysis may be regarded as substitution reactions (via addition-elimination pathways). The activation of (methylthio)methyl esters by S-methylation [51] is an example demonstrating the usefulness of polarity alternation accentuation.

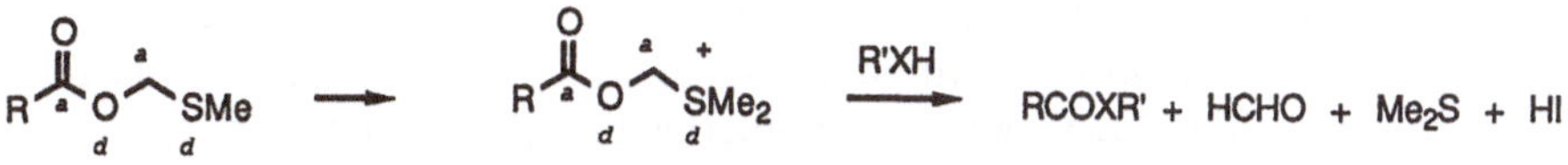

$$ \text{(34)} $$

Acyl triflamides are excellent acylating agents for alcohols and amines [52].

Classical methods for carboxyl activation include formation of anhydrides, either homo- or mixed (e.g. phosphoryl, sulfonyl, etc.). Further activation is possible by adding 4-dimethylaminopyridine as catalyst [53]. The active acylating agents are throught to be

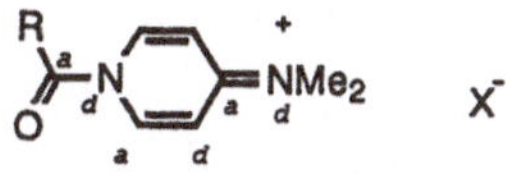

N-Acyloxypyridin-2-thiones derived from aromatic acids undergo thermal or photoinduced decomposition to give anhydrides [54]. Active pyridinium carboxylate intermediates are formed via a bimolecular reaction.

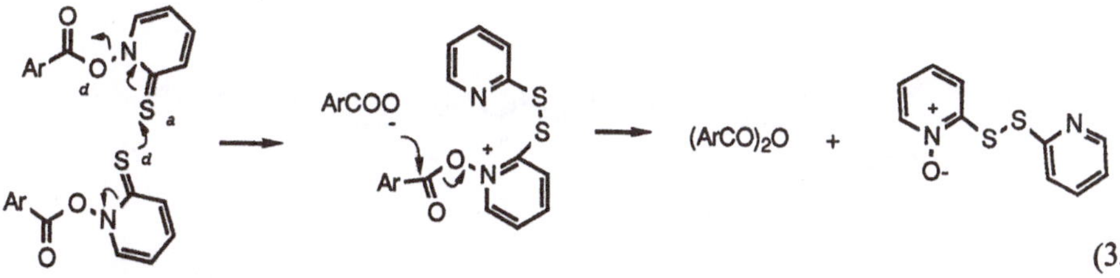

$$ \text{(35)} $$

Extension of polarity alternating sequence in each case is apparent.

EEDQ is a valuable reagent for carboxyl activation in the coupling of amino acids [55]. Ideal *d-a* pairings of various reactive moieties facilitate the formation of the semicarbonate intermediates. Of course, a major driving force for this transformation is the aromatization of the dihydroquinoline system.

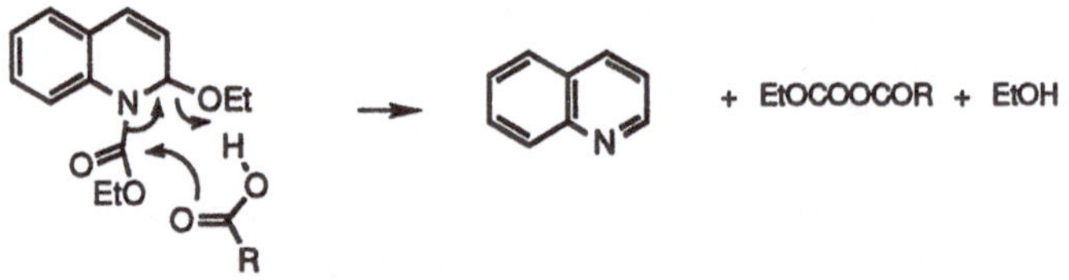

$$
\text{(36)}
$$

Other effective ways of carboxyl activation involves the use of carboxylphosphonium [56] and *O*-acylisourea intermediates [57].

It seems proper to digress here to emphasize the importance of proximity effects which can outweigh electronic factors. Thus, racemization of an amino acid residue during peptide synthesis using carbodiimides is due to the ease of intramolecular C-to-N proton transfer within the *O*-acylisourea intermediate [58]. It should be noted that the proton transfer involves a six-centered transition state in which each atom is related to its neighbors in a polarity alternating sense, except the α-carbon of the amino acid residue. Apparently, the acquiescence of one individual atom to a larger system is energetically much less significant.

$$
\text{(37)}
$$

Alcoholysis of hydrazides [59] and epimerization of the 7α-epimers of cephalosporin analogues [60] can be effected with the aid of chloral. Because of the donor chlorine atoms the imines formed initially are conducive to conjugate dehydrochlorination or prototropic shift. It is likely that acylazo compounds are intermediates for the ester synthesis.

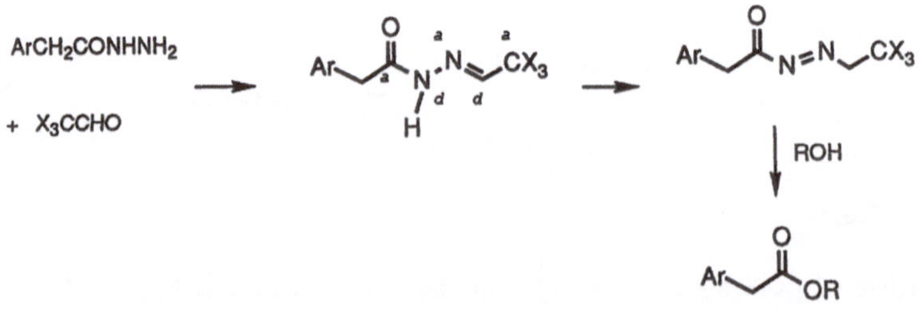

$$
\text{(38)}
$$

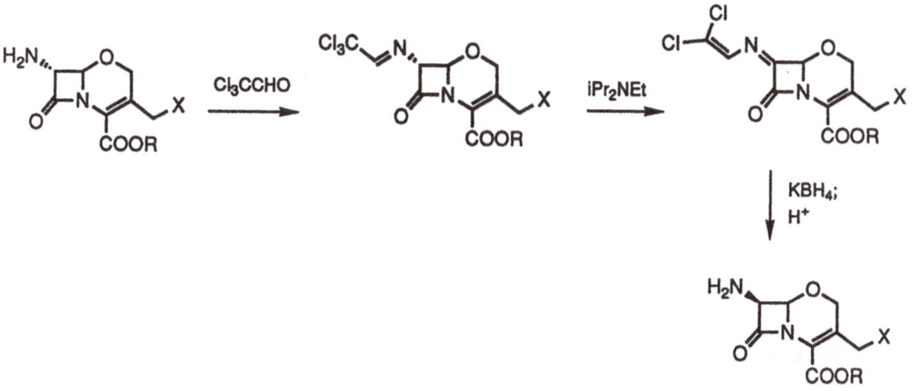

(39)

Hydrolysis of tosylhydrazones has been achieved by treatment with alkaline hypochlorite [61]. This allylic activation ensures the sp^2 hybridized carbon behave as an avid acceptor.

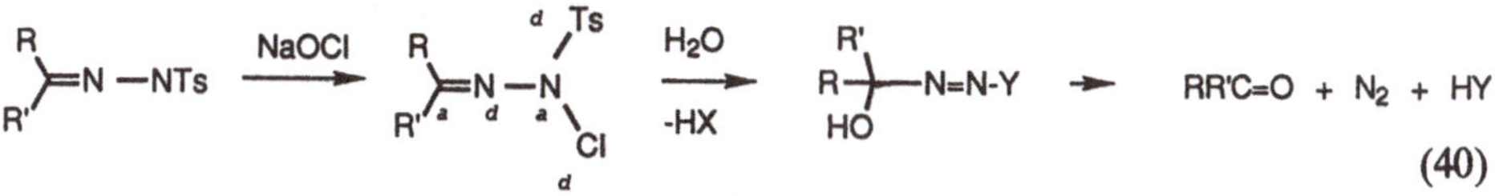

(40)

An S_N2' displacement constitutes the key operation in which the ester group at C-3 of catharanthine is introduced [62]. Thus, a contrapolarizing change of the β-position from d to a allows the desired reaction to be performed, using cyanide ion as the nucleophile.

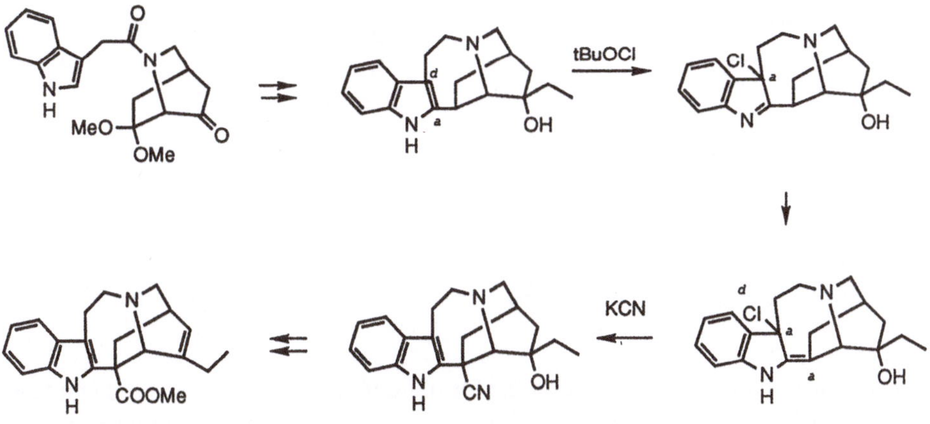

catharanthine

(41)

The following equations depict the intramolecular substitution of an allylic oxy group by a sp-hybridized carbon nucleophile [63]. A silyl group effectively directs its adjacent carbon to neutralize the allyl cation.

95

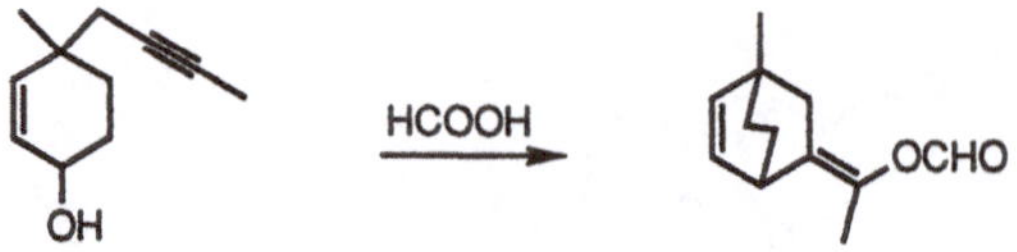

$$(42)$$

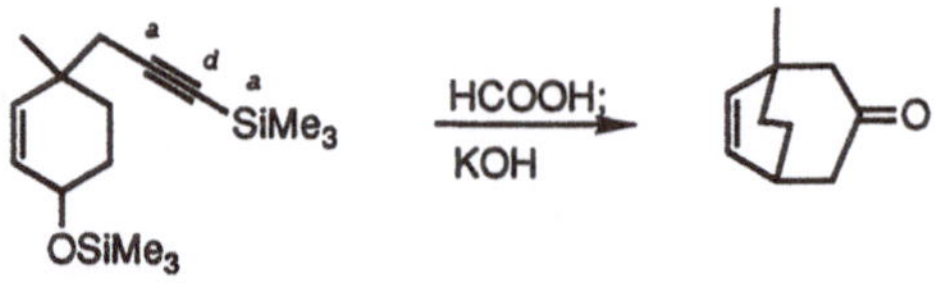

$$(43)$$

Palladium-mediated substitutions of silylated allylic compounds are not subject to steric hindrances. The silicon atom exerts control over the reaction site such that γ-substitution results [64]. While the electrophiles are unavoidably disjoint with respect to the acceptor silicon, one of the two limiting forms, i.e., the α-silylcarbenium ion, is much more unfavorable (*a-a* arrangement) and therefore its population is expectedly low.

$$(44)$$

Arylation of alkenes catalyzed by palladium compounds is known as the Heck reaction [65]. While these reactions are very sensitive to steric effects, subtle electronic contributions to the regiochemical outcome may be assessed by comparing reactions of alkenes with similar substitution patterns. The arylating agents, being the nucleophilic arylpalladium species, tend to dictate the facility of their reaction with unsymmetrical, electron-deficient alkenes.

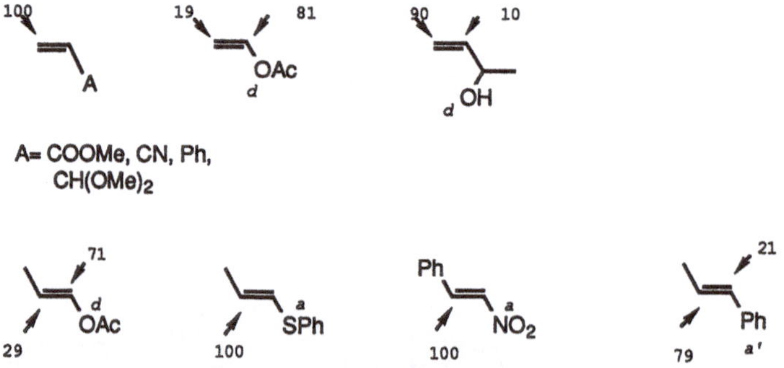

The ratios of attack represented in the formulas [66] reveal a pattern predictable by PAR. The differences between oxy and thio analogues are most intriguing because the sulfur substituent behaves as an acceptor. The dramatic dichotomy may not be purely attributable to a Pd..S coordination in one type of compounds.

Arylation of allylic alcohols leads to 1-aryl-3-alkanones. The orientation for this reaction is in compliance with the general trend indicated above. Furthermore, the accepted reaction mechanism points to a "Pd-H" elimination process in which H departs as a hydride. The generation of enol intermediates is facilitated by the donor hydroxyl group.

$$\underset{d}{Ar}\text{-}\underset{a}{Pd}X \quad + \quad \overset{a}{\diagup}\!\!\diagdown\underset{d}{\underset{|}{OH}}\!\!\diagdown R \quad \longrightarrow \quad Ar\diagdown\!\!\diagup\overset{PdX}{\underset{d}{\underset{OH}{\overset{|}{\diagdown}}}}\!\!\overset{d}{\underset{R}{\overset{H}{\diagup}}} R \quad \longrightarrow \quad Ar\diagdown\!\!\diagup\overset{R}{\underset{OH}{\diagdown}} \quad \longrightarrow \quad Ar\diagdown\!\!\diagup\diagdown\overset{R}{\underset{O}{\diagdown}}$$

(45)

It is interesting to note that the fluoride ion-promoted Si-C bond cleavage of vinyl silanes is greatly facilitated by the presence of a hydroxyl group at the tetrahedral β-carbon atom [67].

$$\overset{Me_3Si}{\diagdown}\!\!\diagup\overset{OH}{\underset{R'}{\overset{|}{\diagdown}}R} \quad \xrightarrow[\text{H}_2\text{O}]{\text{F}^-;} \quad \overset{H}{\diagdown}\!\!\diagup\overset{OH}{\underset{R'}{\overset{|}{\diagdown}}R}$$

(46)

Many factors contribute to the success of the desilylation, including a six-centered transition state in which F^- associates with both Si and the proton of the hydroxyl group. Perhaps the electronic influence by the oxygen atom on the silylated sp^2-carbon is such that a better donor for the latter atom results.

3 Eliminations and Fragmentations

Arynes are formed when haloarenes are treated with a strong base. Deprotonation at the ortho position of the halogen atom is favored because of polarity alternation, and the facility of benzyne formation from halobenzene (PhF > PhCl > PhBr ≫ PhI) also corresponds with the relative strength of the donor.

$$\underset{\overset{|}{H}\,^a}{\overset{\overset{a}{X}\,_d}{\bigcirc}} \quad \xrightarrow{\text{base}} \quad \bigcirc$$

(47)

Reversal of the E2 leaving group ordering has also been observed in the alicyclic series, i.e. *syn* elimination of *trans*-1,2-dihalocycloalkanes [68]. This reactivity pattern is very well acounted for by considering the polarity influences. In general, the halogen atom with a highly acceptor character (I > Br > Cl > F) would loosen its geminal hydrogen toward a donor reagent, and contrarily, the highly donor congener (esp. F) renders its geminal hydrogen more difficult to be abstracted by bases (donors).

$$\underset{\overset{}{\underset{Na}{\diagdown}\!\!\!-\!\!\!\underset{NH_2}{}}}{\overset{H}{\diagdown}\!\!\!\bigcirc\!\!\!\overset{Br}{\diagup}} \quad \xrightarrow[\text{tBuONa}]{\text{NaNH}_2} \quad \bigcirc\!\!\!\overset{}{\underset{Br}{\diagdown}}$$

(48)

Dehydrohalogenation of 2,3-dihalo-2,3-dihydrobenzofurans gives 3-halobenzofurans [69]. Rate measurements have indicated higher difficulties in deprotonation at the benzylic position when the benzylic halogen is highly donative (i.e. F ≫ Cl > Br). The trend is evident for both *syn*-E2 and $E1_{cBI}$ mechanisms.

The thermal *syn*-elimination reactions of X-oxides have assumed great importance in synthesis because of their mildness and regioselectivity.

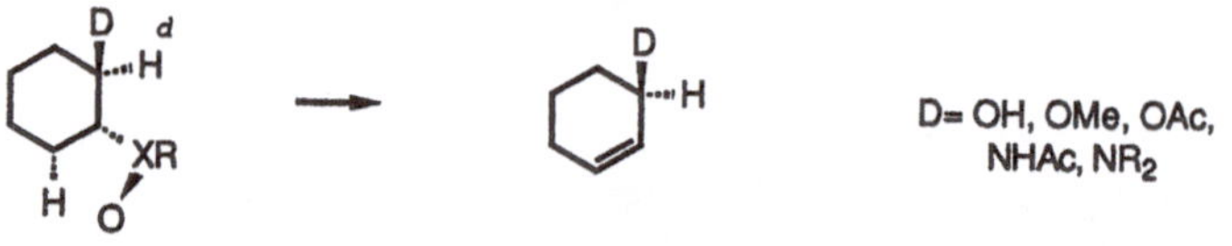

D= OH, OMe, OAc,
NHAc, NR2
(49)

A= COR, CN, NO2,
SOAr, SO2Ar
(50)

It is apparent that a *vicinal, trans* donor subtituent favors the formation of an allylic product, whereas an acceptor group strongly biases the generation of a vinylic derivative. The regioselectivity is a result of the electronic transmission by the polar group in the same manner as discussed above [70].

The extraordinary dehydrofluorination of 1,1,1,3,3,3-hexafluoro-2-propanol [71] on exposure to *n*-butyllithium shows an overwhelming effect by the fluorine atoms on the carbinolic center. In other words, the influence of the single oxygen is largely suppressed.

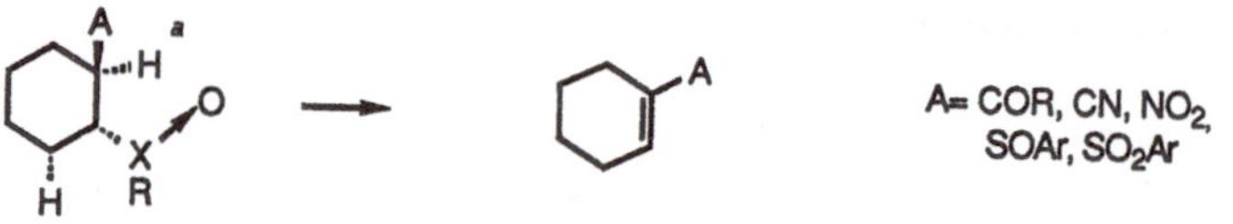

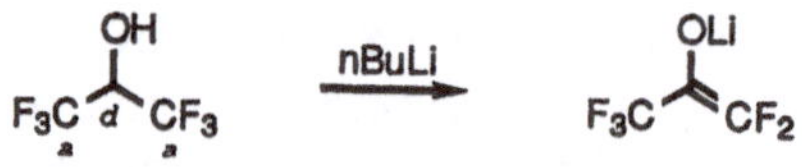

OH
F3C d CF3
nBuLi
OLi
F3C CF2
(51)

While no convincing reasons have been put forward to explain the different behavior of the two photocycloadducts derived from 5-fluorouracil, the elimination process is definitely favored by the presence of a methoxy group, and even more so with a *gem*-dimethoxy functions [72]. Such a methoxy donor would confer acceptor character to the angular hydrogen.

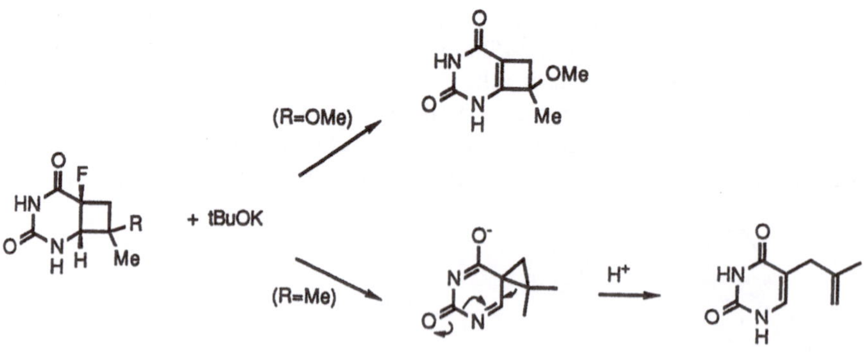

+ tBuOK
(R=OMe)
(R=Me)
OMe
Me
H+
(52)

The solvolysis of 2-trimethylsilylcyclohexyl trifluoroacetates proceeds with enormous rate enhancements over the nonsilylated substrates, especially when the *vicinal* heterosubstituents are trans [73]. This β-silicon effect is a special case of activation by polarity alternation.

Oxidation of alkyl halides via reaction with N'-triflyl hydrazides exploits the facile elimination of triflinic acid [74]. Acidification of N-H and C-H by a β-donor is essential to the elimination and the prototropic shift.

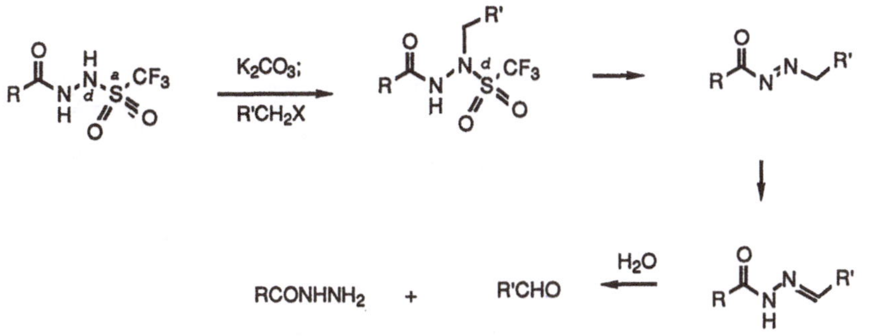

$$RCONHNH_2 \quad + \quad R'CHO \quad \xleftarrow{\quad H_2O \quad} \quad (53)$$

Regeneration of carbonyl compounds from 5-(2-pyridyl)-1,3-dioxanes [75] can be effected by N-methylation and treatment with a mild base. The picolyl carbon is a donor site, and the loss of a proton from that position is facile.

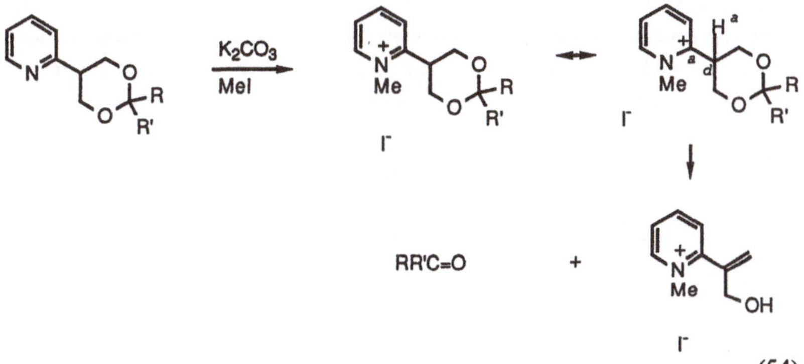

$$RR'C=O \quad + \quad (54)$$

The following equations show the difference a strategically located ester group can make in the direction of aziridine opening [76]. The ester favors deprotonation at the conjoint benzylic position, leading to the benzazepine as the major product.

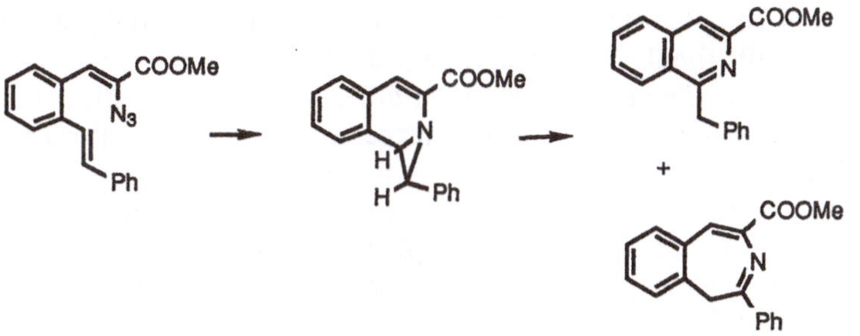

$$(55)$$

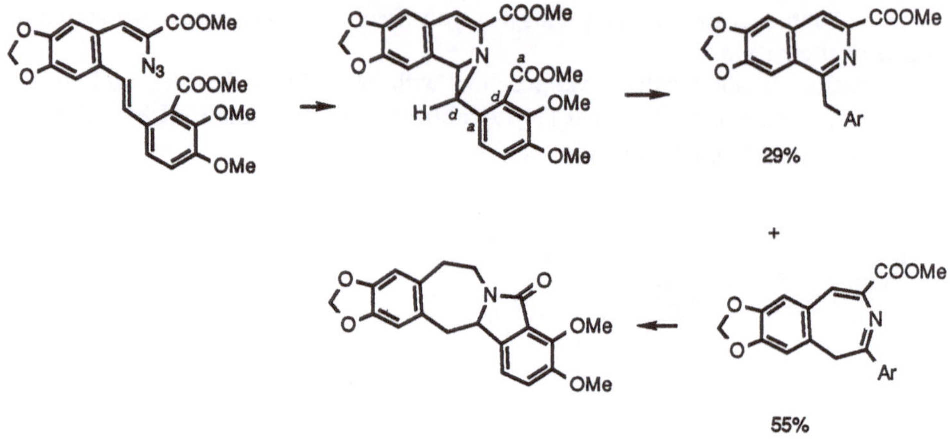

(56)

An intriguing reaction dichotomy arising from variation of the aryl group in *N*-arenesulfonyltriazenes [77] may be the consequences of polarity alternation. The subtle difference of the aromatic substituent can affect the acidity of the methine hydrogen. For example, the *p*-nitro group acidifies the methine very strongly, making a facile tautomerization of the diazo form to the hydrazo form. On the other hand, change of the aromatic moiety to a *p*-tolyl or trisyl group might be sufficient to impede the tautomerization such that elimination of the arenesulfinic acid can compete successfully. The end results are the preferential formation of diazo compounds and azido compounds, respectively, as indicated in the equations.

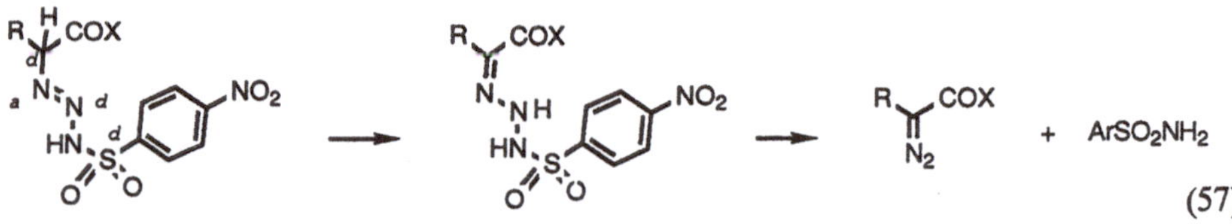

(57)

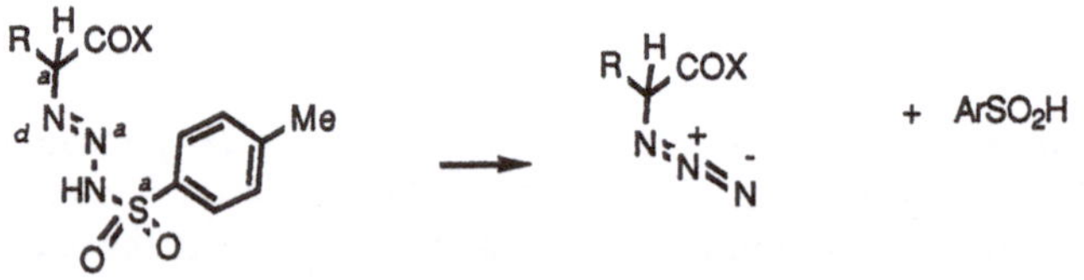

(58)

It should be noted that according to the distribution of the polar functionalities opposite results should have been obtained. However, a faster tautomerization than fragmentation in the nisyl system and a difference in mechanistic details for the two series might be responsible for the 'unexpected' observations. Perhaps a spirocyclic intermediate is involved in the fragmentation of the nisyl derivative.

The major chloroform extractible product from the decomposition of 1-deoxy-1-dibenzylamino-D-fructuronic acid in pH 6.0 buffer is 4-hydroxy-5-methyl-3(2*H*)-furanone [78]. Treatment of the same Amadori compound in 2N sulfuric acid gives 2-furaldehyde [79].

The divergent pathways arise from selective enolization. Thus, in media of lower acidity enolization occurs toward C-3, whereas in a stronger acid complete N-protonation shifts the enolization toward C-1. The amino donor confers certain acceptor character C-1 (although this is a captodative situation), but upon protonation it is converted into an acceptor and its adjacent carbon atom, a donor.

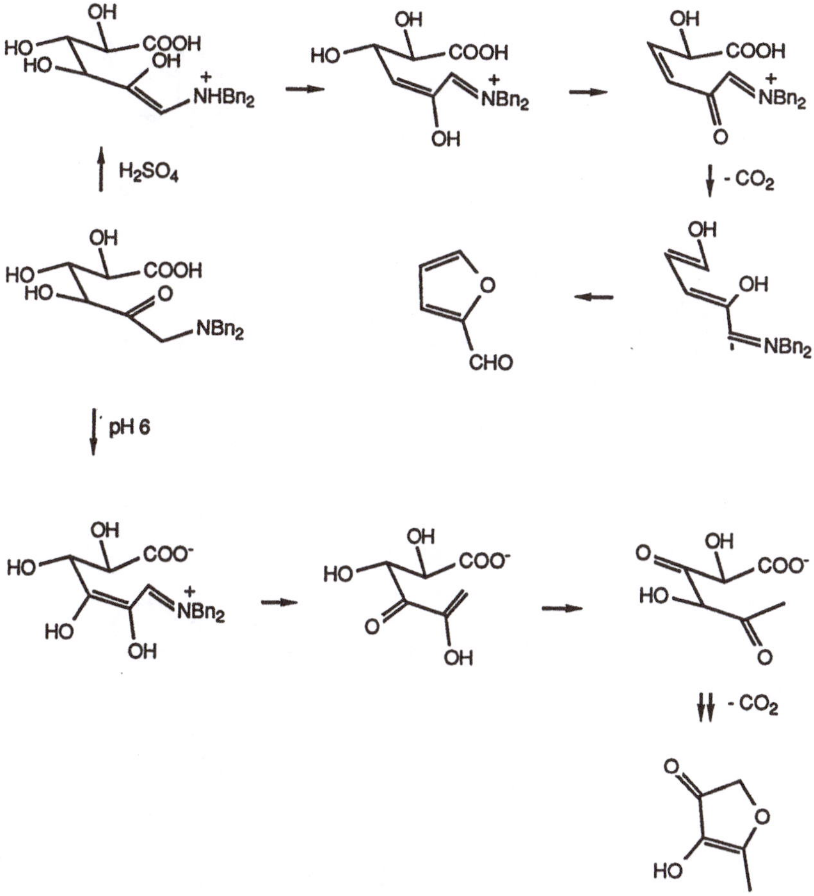

(59)

Compounds in which two donor atoms are linked by a three-carbon chain undergo C-C bond cleavage readily. Well-known reactions are the retro-aldolization, retro Claisen, retro-Michael, and retro-Mannich reactions. Significant application of such processes to synthesis of complex natural products include approaches to caryophyllene [80], nootkatone [81], trihydroxydecipiadiene [82], hybridalactone [83], and mesembrine [84].

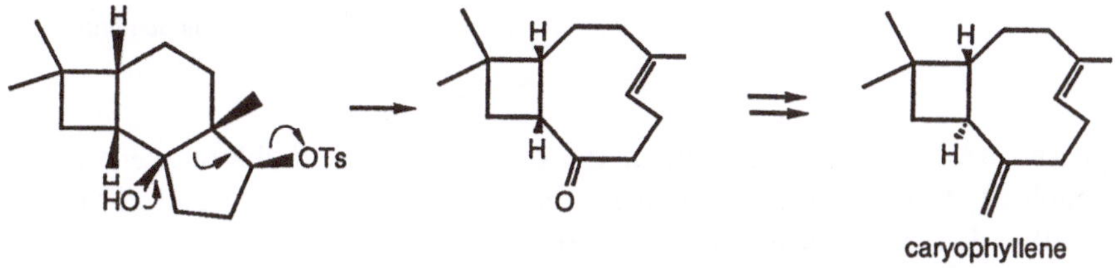

caryophyllene (60)

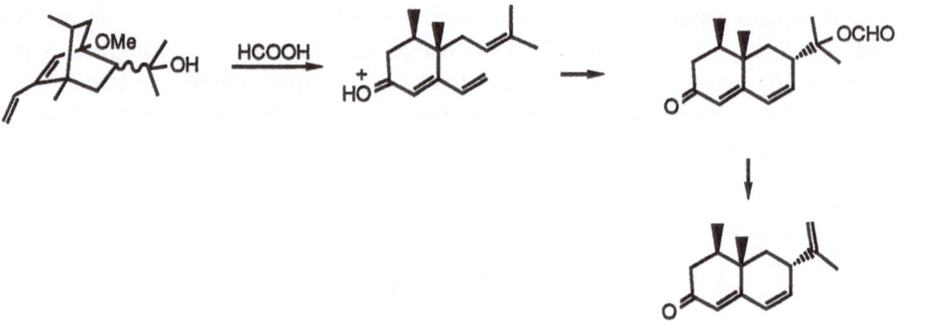

nootkatone (61)

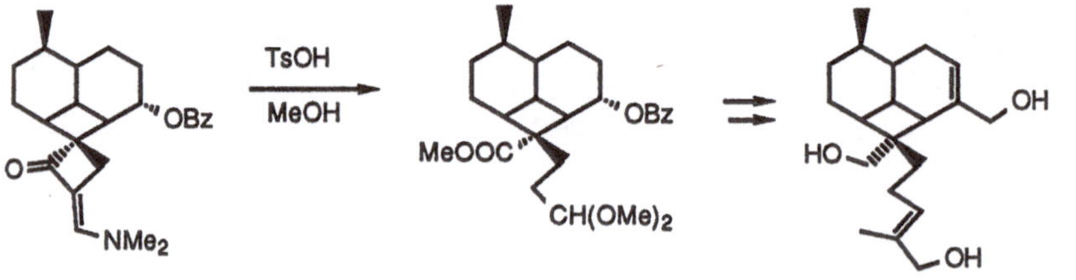

trihydroxydecipiadiene (62)

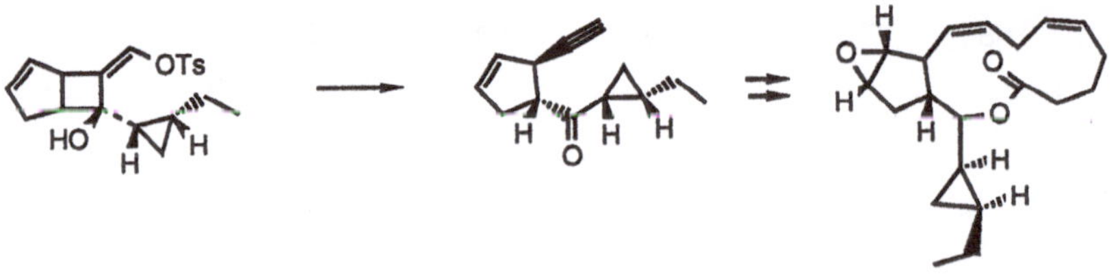

hybridalactone (63)

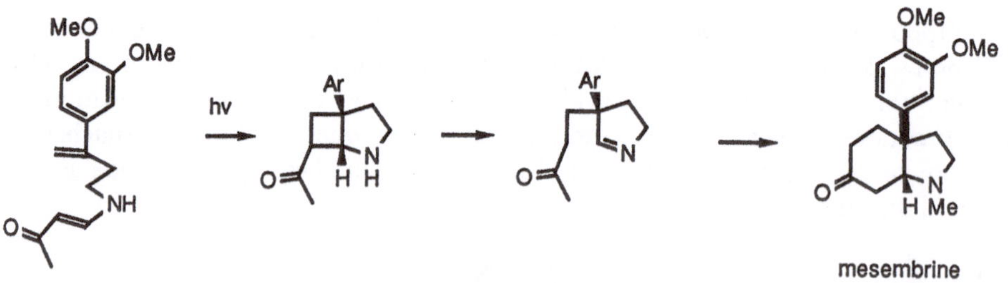

mesembrine

(64)

The base-catalyzed rearrangement of the cyclopentadiene-dichloroketene adduct gives tropolone. The reaction course passes through an S_N2' displacement of the enol, fragmentative dehydrochlorination, and enolization [85].

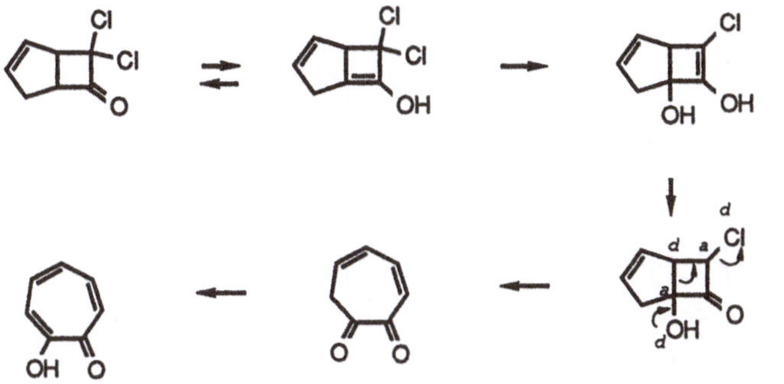

(65)

In α-nitrocycloalkanones the donor oxygens are also separated by three atoms. Consequently, these compounds can also undergo fragmentation. A synthetically useful method based on the enhanced donor activity of the α-carbon and the fragmentability is a macrolactone formation. Particularly interesting is the secoannulation onto a benzene ring by the ω-nitroacyl element via a Michael reaction with benzoquinone [86]. It is noteworthy that in the Michael adduct the ketol subunit is disjoint through switching (the proximal OH is not derived from the enone acceptor). Because benzoquinone contains two disjoint carbonyl functions, reactions of such molecules in the normal sense (i.e. without contrapolarization) always result in a pair of functional groups in disjoint arrangement.

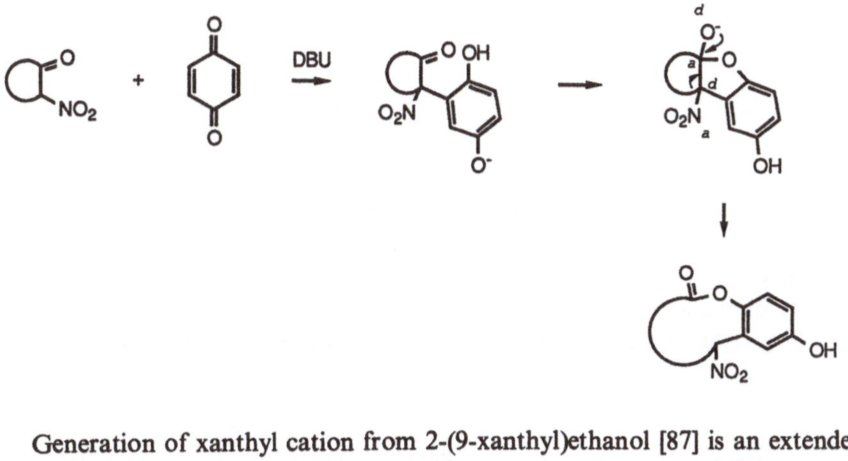

(66)

Generation of xanthyl cation from 2-(9-xanthyl)ethanol [87] is an extended Grob fragmentation. The intervening chain of separating the terminal donor groups may also incorporate heteroatoms, as shown in the deconvolution of a decalindione monoxime tosylate [88].

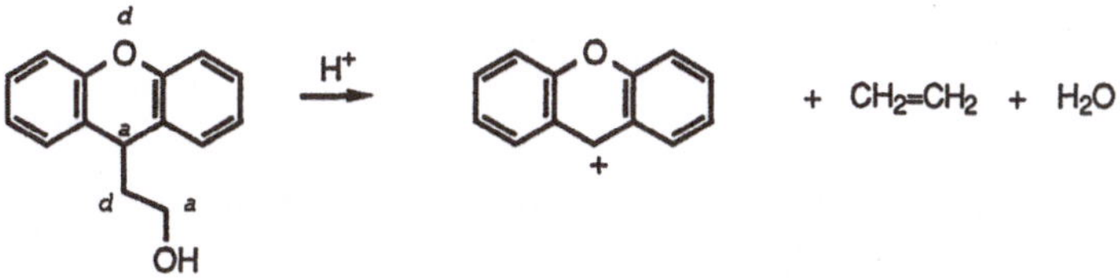

(67)

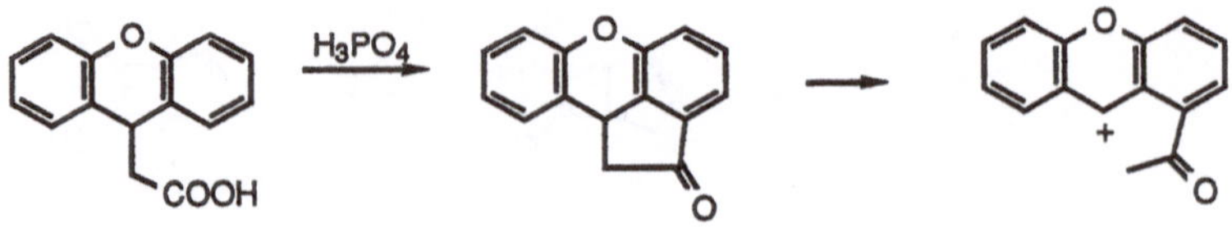

(68)

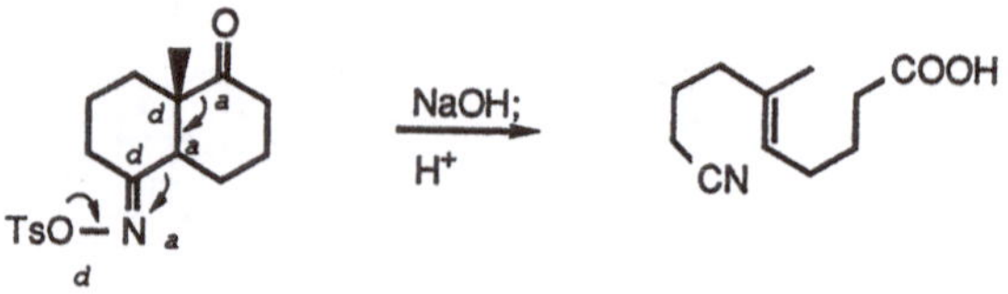

(69)

Noteworthy is a thermal cycloreversion of a tricyclic tosylate carboxylate that translates chirality of five carbon centers into three double bonds (2C=C and 1C=O) [89].

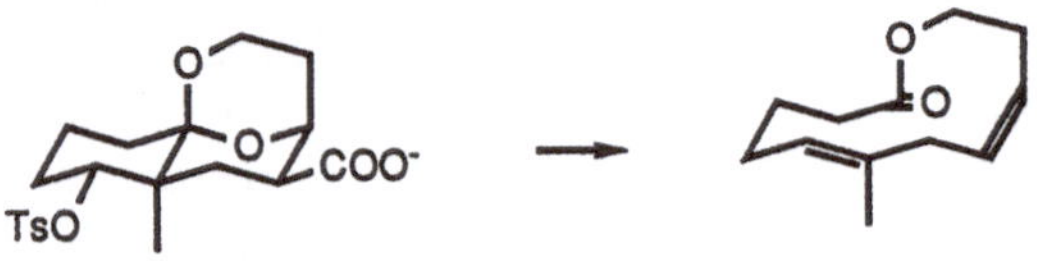

(70)

Decarboxylation of chloroformic esters is greatly catalyzed by dimethylformamide [90]. As indicated in the following equation, an iminium intermediate with contiguous *d-a* pairs is involved.

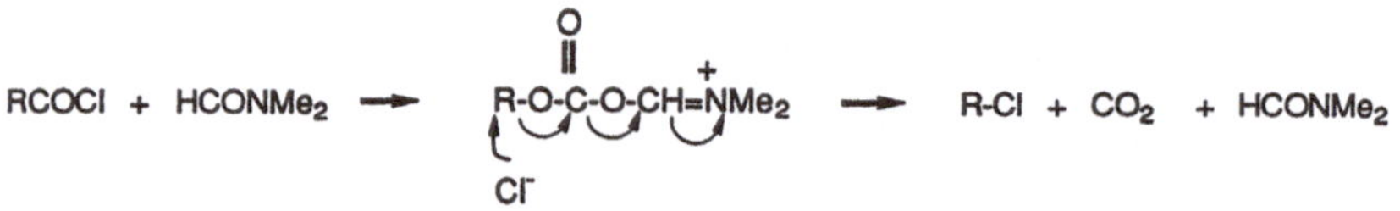

(71)

Acetals are acid sensitive but alkaline stable because their hydrolysis requires activation of the oxygen atom by protonation or coordination with a Lewis acid. An ingenious way of differentiating two 2-methoxytetrahydrofuran moieties during a synthesis of bilobalide [91] is to treat the tetracyclic intermediate with potassium hydroxide. Expulsion of a methoxide ion is triggered by attack of the γ-lactone ring five carbons away.

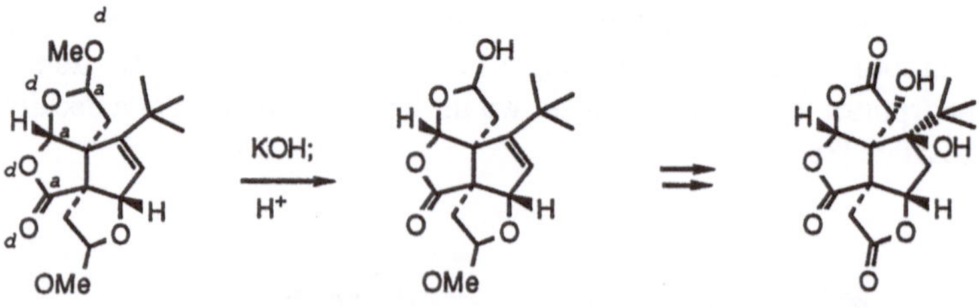

bilobalide   (72)

Assembly of the indolazepine intermediate for synthesis of vincadifformine [92] encompasses a 1,2-rearrangement-retro Mannich-Mannich reaction sequence. Needless to say, the smooth transformation is the result of a perfect polarity alternation set-up.

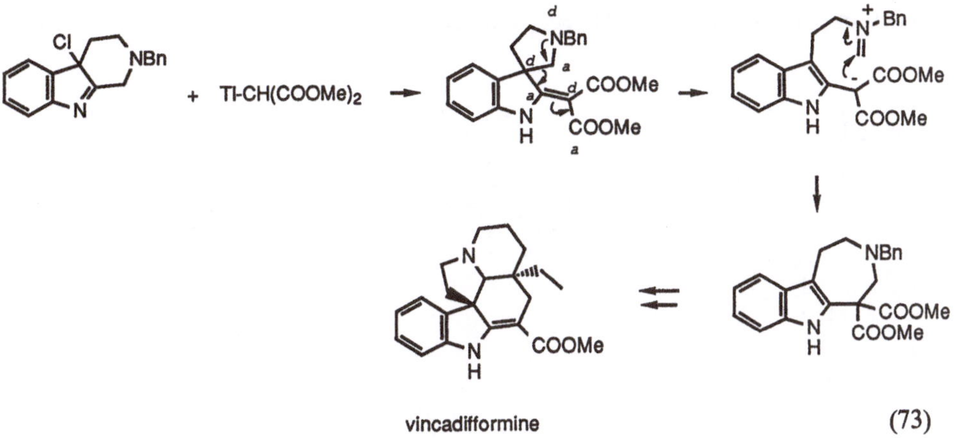

vincadifformine

(73)

The 9-membered ring present in quebrachamine has also been established using the fragmentation strategy [93].

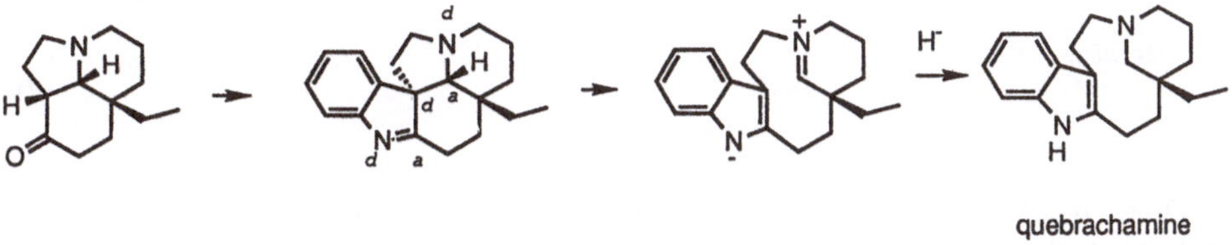

quebrachamine

(74)

Aspidospermidine oxime undergoes double skeletal fragmentation to give a tri-cyclic iminium ion which cyclizes at the α-position of the indole nucleus [94].

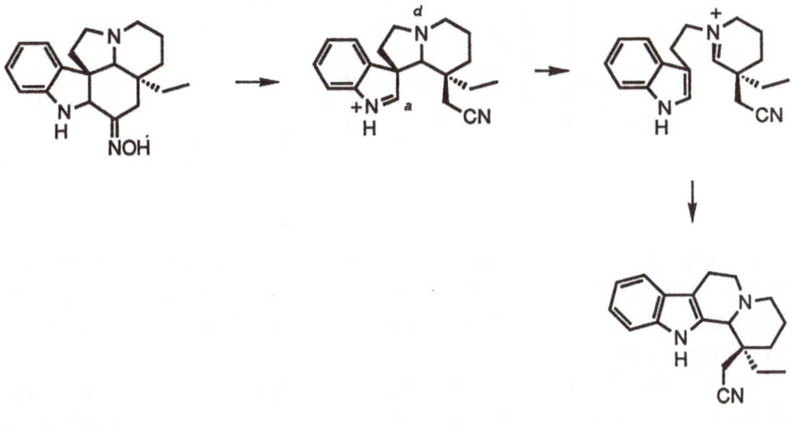

(75)

105

An even more complex series of bond formation and cleavage attends the construction of a secodine precursor for tabersonine [95]. The initial tetracyclic intermediate provides two 3-carbon links between the two nitrogen atoms and therefore reaction pathways for skeletal reorganization.

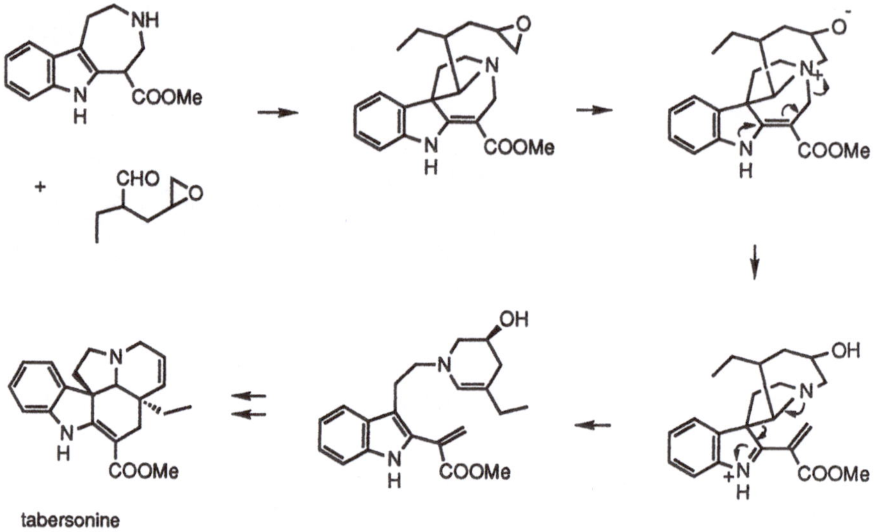

tabersonine

(76)

There is an intriguing observation concerning the equilibratability of certain indolyltetrahydropyridine. The N_b-unsubstituted isomers undergo equilibration readily, yet the N_b-hydroxyethyl derivatives are stable [96].

The simplest explanation seems to be that the hydroxyl group makes the nitrogen atom a poorer donor. Naturally other factors such as hydrogen bondings that render the molecules unfavorable to fragment must also be considered.

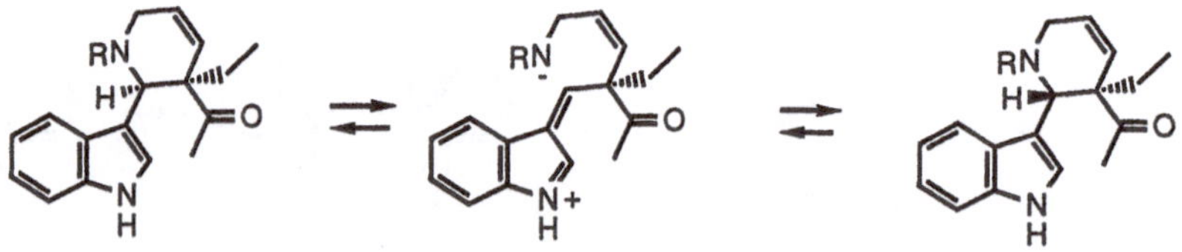

(77)

Thermolysis of certain vinylic azides gives rise to cyano compounds. Representatives of this interesting class of reactions (zwittazido cleavage) [97] are outlined in the following. The fragmentation is predicated by a β-atom which can assume an acceptor role readily.

In the zwittazido cleavage substituents of opposite polarity are present at each end of the C-C bond to be broken. This structural pattern also exists in the α-diketone monothioketals [98], and naturally, fragmentation occurs readily in those compounds.

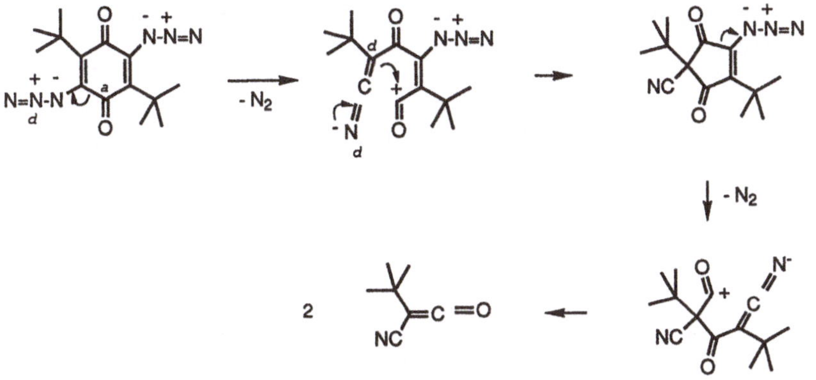

(78)

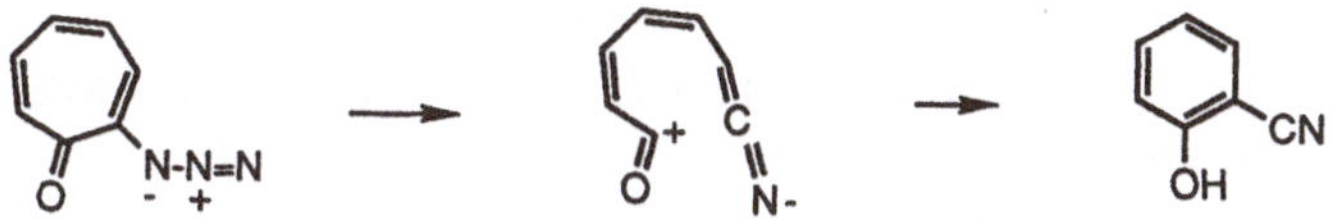

(79)

(80)

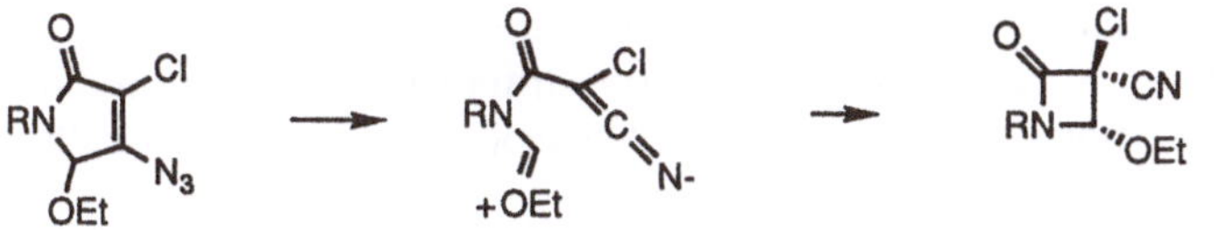

(81)

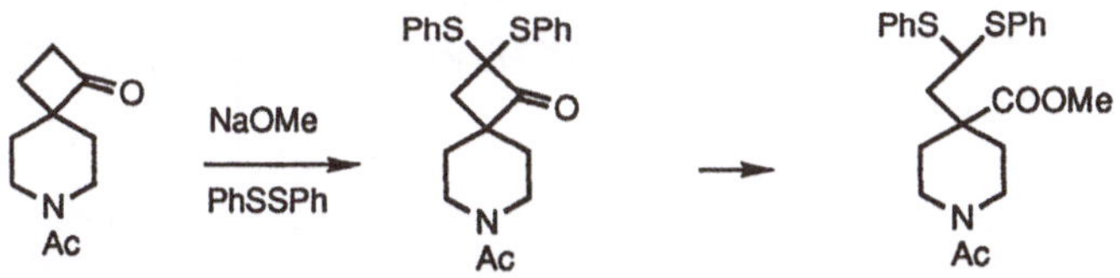

(82)

Another example is the indirect acetonylation of a 1,2,3,4-tetrahydropyridine derivative at the β-position of the enamide system [99]. The reaction involves a double alkylation to furnish an acetylcyclopropane which undergoes fragmentation on acid treatment. It is noted that polarity alternation and ring strain relief allow the facile and regiospecific C-C bond cleavage to proceed, the intervention of a cyclopropane intermediate also leads to disjoint system. This is a fundamental strategy [100] which has been frequently employed to gain access to compounds with disjoint function-alities.

Such disjoint molecules may serve as synthetic intermediates. Thus a [3 + 2] annulation method exploits the facile ring opening of cyclopropanols and the transfer of the donor site to the β carbon of the resulting ketones, the nascent carbanions may be intercepted by a vinylphosphonium species and an intramolecular Wittig reaction follows [101].

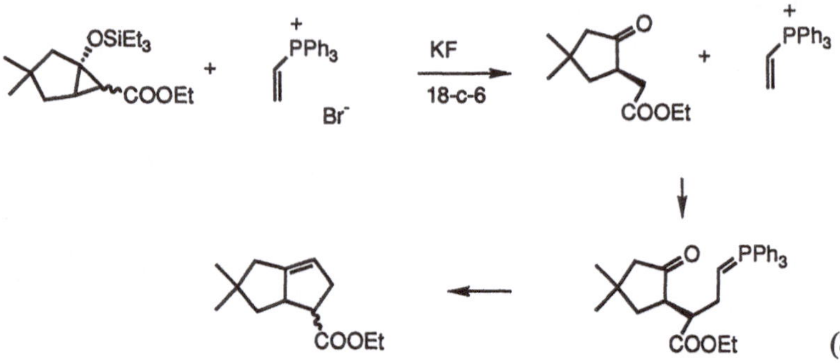

(83)

A variation at the final stage of the carbanion trapping to a Michael addition provides an entry to the octalone system. A synthon for dihydrocompactin is readily assembled [102].

(84)

A reaction of bromobenzene with ethyl malonate in the presence of sodamide affords homophthalimide [103]. This fascinating reaction involves sequential elimination, addition, intramolecular acylation, fragmentation, and addition. Excellent polarity alternation along the chains of atoms in various intermediates allows so many steps to proceed efficiently.

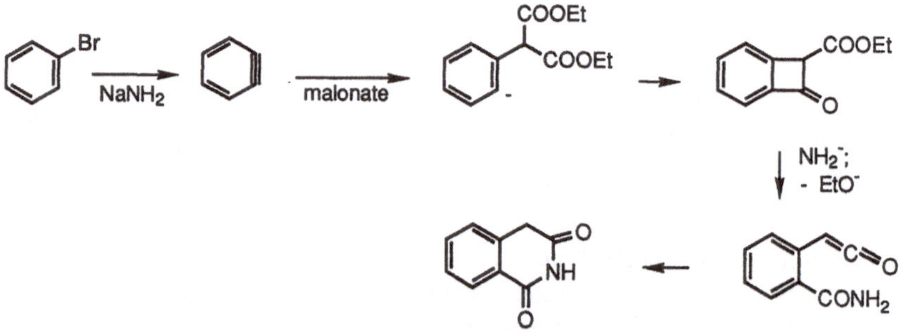

(85)

Fragmentation of a disjointly substituted carbon chain requires contrapolarizing operation such as oxidation. A method for synthesizing macrocyclic lactones is illustrated below [104]. Introduction of an oxyalkyl chain follows the conjugate addition of a stannyl anion to a cyclic enone results in a bicyclic lactol Exposure of the lactol to lead tetraacetate initiates oxidation at the hydroxyl group and subsequent fragmentation. Thus, all four skeletal atoms constituting the original enone system undergo contrapolarization in this step.

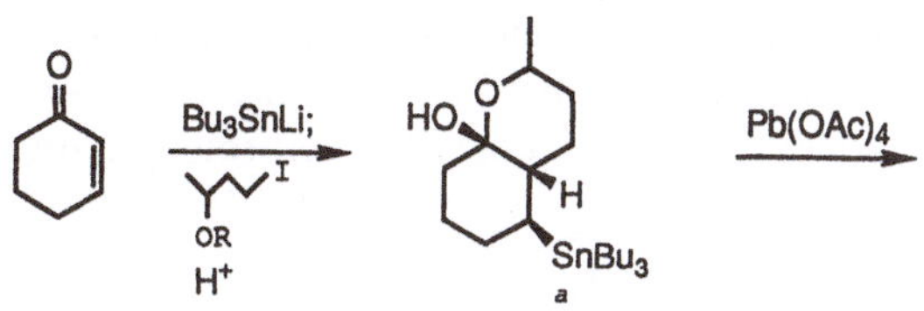

$$\text{(86)}$$

4 Additions and Cycloadditions

4.1 Additions to Multiple Bonds

Additions to carbon-carbon multiple bonds initiated by electrophiles are generally governed by the Markovnikov rule. However, the rule must be modified to accommodate such substrates as vinylsilanes. The so-called anti-Markovnikov hydrohalogenation [105] is to be contrasted to the Markovnikov addition for allylsilanes. In fact, when one recognizes the acceptor role of the silicon atom and applying the polarity alternation rule, the puzzling results become self-consistent.

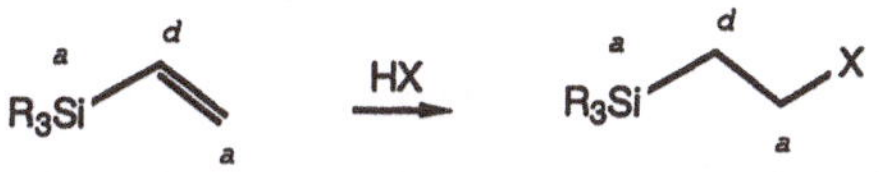

$$\text{(87)}$$

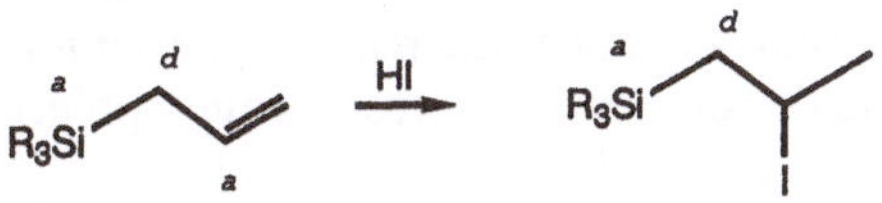

$$\text{(88)}$$

The directing effect of an allylic silicon atom is also manifested in hydroboration [106]. This exquisite control is extendable to tetrasubstituted olefins [107] making the access of *trans*-a-decalones quite readily.

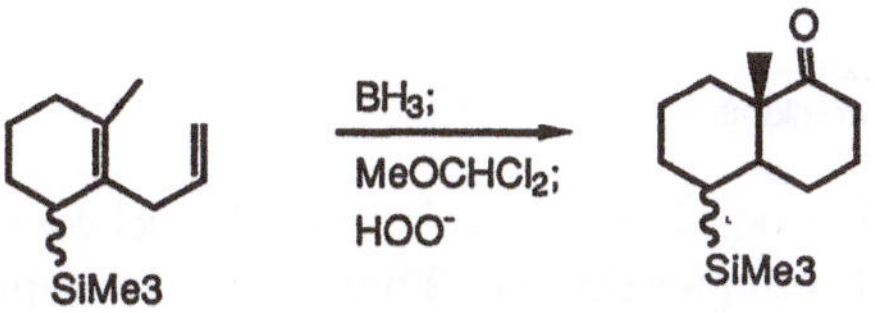

$$\text{(89)}$$

As expected, allylic alcohols and ethers afford *vic*-diol derivatives upon hydroboration. However, in the presence of Wilkinson's catalyst, a reverse regiochemistry for the hydroboration is observed [108]. There is the possibility that boron, rather than hydrogen, is transferred from the metal to the olefin. In other words, the boron acts as a donor.

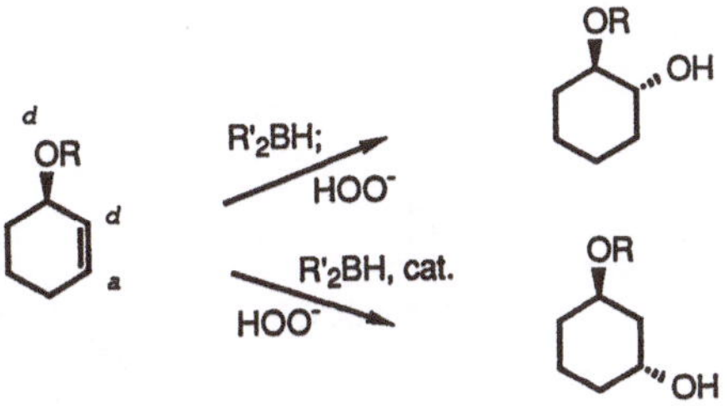

$$\text{(90)}$$

109

The regiochemistry of the addition of phenylselenenyl chloride to allylic alcohols and their esters [109] can simply be rationalized by using the polarity alternation rule except in those cases where steric factors become the controlling parameters.

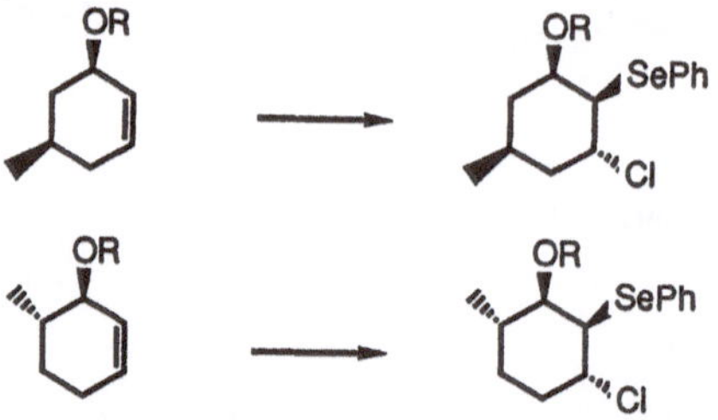

1,2-Disubstituted alkenes undergo iodohydroxylation regioselectively when the allylic position bears a hydroxyl group [110]. The attack on the distal carbon of the iodonium intermediates is favored on the polarity alternation basis.

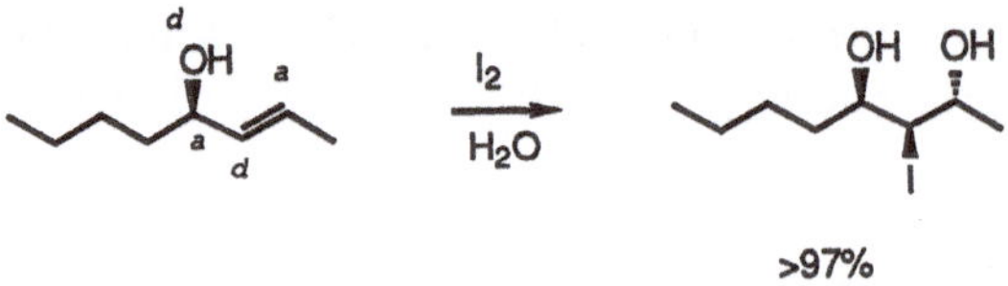

$$(93)$$

>97%

Nitrocyclohexene is rather unreactive toward many nucleophiles. Interestingly, an oxygen substituent at C-6 substantially enhances the reactivity, even if the substituent is non-leaving [111]. Evidently the donor exerts its effect through polarity alternation accentuation.

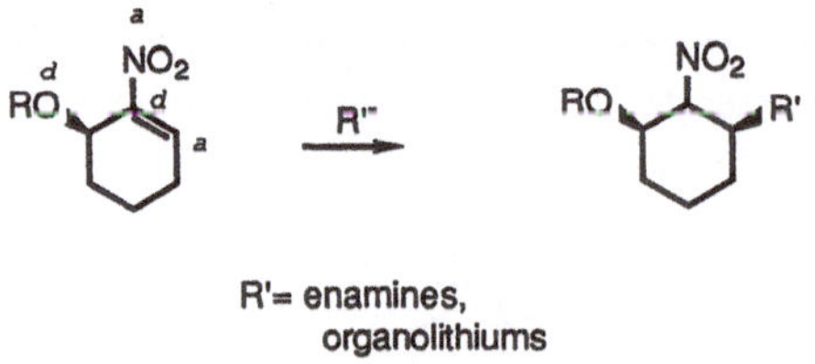

$$(94)$$

R'= enamines,
organolithiums

Functionalization of the double bond of 2-azabicyclo[2.2.2]oct-5-ene derivatives has been shown to be remarkably regioselective. Thus, oxymercuration [112] and selenohalogenation [113] lead to adducts which are convenient precursors of iso-quinuclidin-5-one and -6-one, respectively. This means an excellent control of the polarity at the double bond termini by the nitrogen atom (Note that simple allylic urethanes also undergo regioselective oxymercuration).

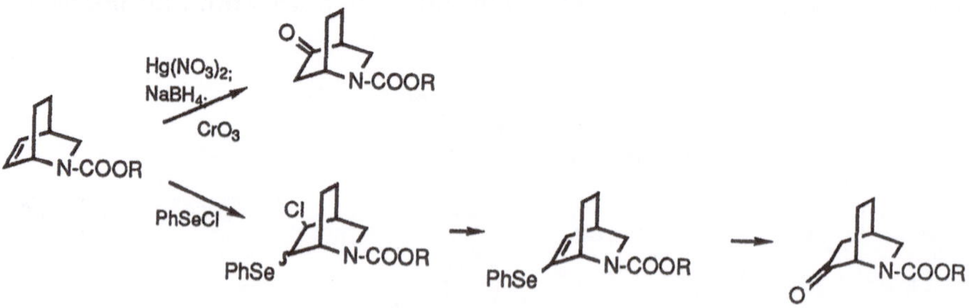

$$(95)$$

Remote function control of the regiochemistry for 1,2-addition in the bridged bicycloalkene systems [114] has been delineated. Again, the polarity alternation rule proves very useful for rationalizing these results.

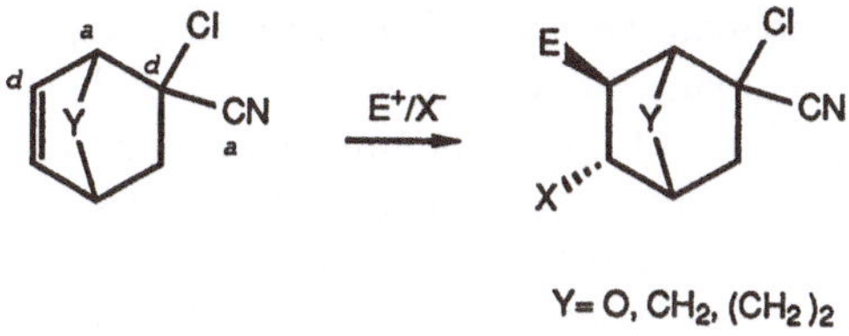

(96)

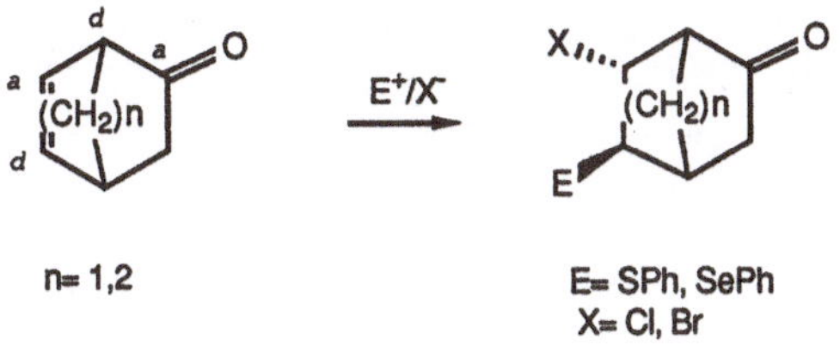

Y= O, CH₂, (CH₂)₂

(97)

n= 1,2

E= SPh, SePh
X= Cl, Br

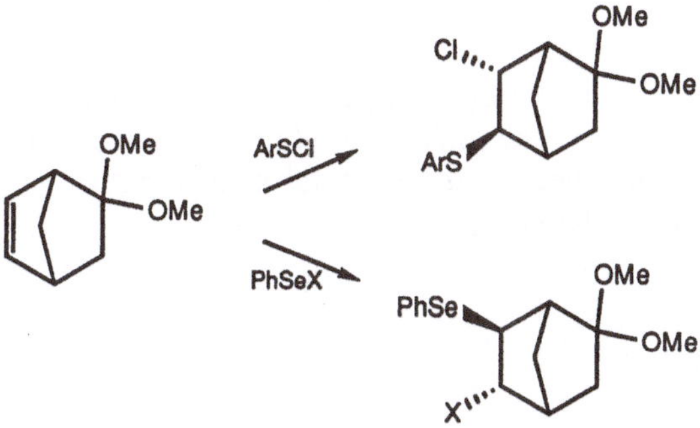

(98)

It should be noted that the last reaction is anomalous owing to the soft nature of the selenium such that the intermediate is an episelenonium species (i.e. less carbocationic) and the attack of the counterion is subject to steric effects such as the 2,6-*endo* interaction.

The first total synthesis of morphine [115] involves hydration of a tetracyclic intermediate. The attack of water from the more hindered face of the double bond could hardly have been predicted. However, the donor nitrogen atom of the molecule may have been responsible for the protonation at C-7 in favor of C-6 via a polarity alternation mechanism.

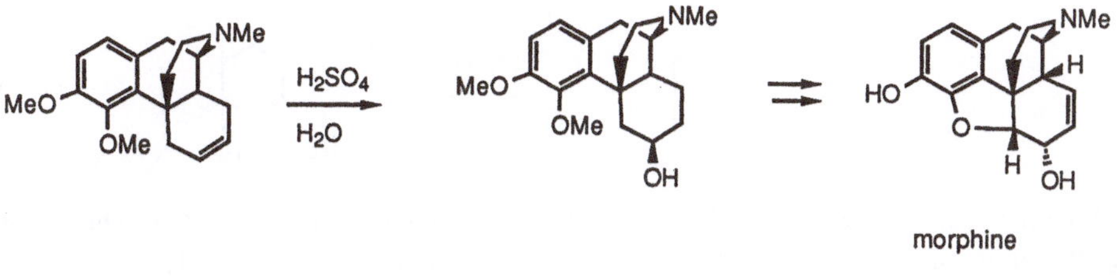

morphine

(99)

111

As expected, additions to alkynes are also subject to control by polar functions nearby. Thus, nucleophilic attack on trifluoromethylacetylene [116], cyanoacetylene [117], and ethylthioacetylene [118] occurs at the terminal *sp*-hybridized carbon atom, the substituent at the other end of the triple being an acceptor in all cases. This behavior is to be contrasted with the mode of addition on ethoxyacetylene [119], aminoacetylenes [120]. The vinylogue, 4-dimethyl-aminobut-3-en-1-yne [121], reacts with aniline at the internal position of the akyne linkage. However, a 2:1 regioselectivity, favoring the methanol adduct predicted by the polarity alternation rule, has been observed for the addition of *N*,*N*-bis(trifluoromethyl)ethynylamine [122]. It is not known whether steric factors play a role in the decreased regioselectivity.

$$
\overset{a\quad d\ a}{F_3C-C\equiv CH} \qquad \overset{a\quad d\ a}{NC-C\equiv CH} \qquad \overset{a\quad d\ a}{EtS-C\equiv CH} \qquad \overset{d\quad a\ d}{EtO-C\equiv CH}
$$

$$
\overset{d\quad a\ d}{R_2N-C\equiv CH} \qquad\qquad \overset{d\quad a\ d}{R_2N-C\equiv CH}
$$

R= F₃C   56%   28%

The notion that silyl groups and alkyl groups belong to different categories in the polarity sense is now established. The cyclization of an iminium salt by alkyne participation is dramatically controlled by the terminal substituent of the acetylene in being a methyl or a trimethylsilyl group [123].

(100)

(101)

The regiocontrol furnished by a silyl group is also revealed in a radical cyclization. Alkyl radicals are generally nucleophilic and therefore may be considered as donors, although not to the same degree as carbanions. In the formation of a β-agarofuran precursor [124] the terminal trimethylsilyl group serves a dual purpose of regiocontrol contribution and suppression of free radical chain polymerization.

(102)

A trimethylsilylethynyl moiety can serve as a vinyl cation equivalent, as shown by an intramolecular reaction with a dienamine [125]. The silver ion catalyst acts as an accentuating agent.

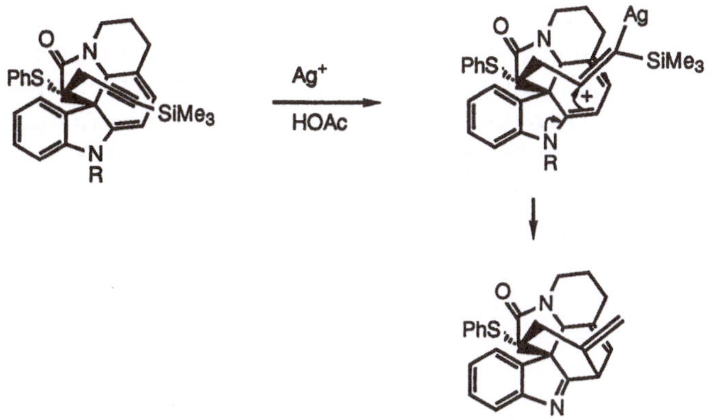

(103)

Polyene cyclization in terpene and steroid synthesis is critically dependent on the terminator in order to generate useful functionalities for further modification of the products. Allyl- and propargylsilanes have proven their value in facilitation of the cyclization and generation of an exocyclic methylene and allene, respectively. Thus, a concise approach to albicanyl acetate [126] and the rapid construction of a tetracyclic precursor of steroids [127] are sufficient to demonstrate the concept. Again, a comparison of the substrates with a silyl group with those having a simple alkyl moiety is very enlightening.

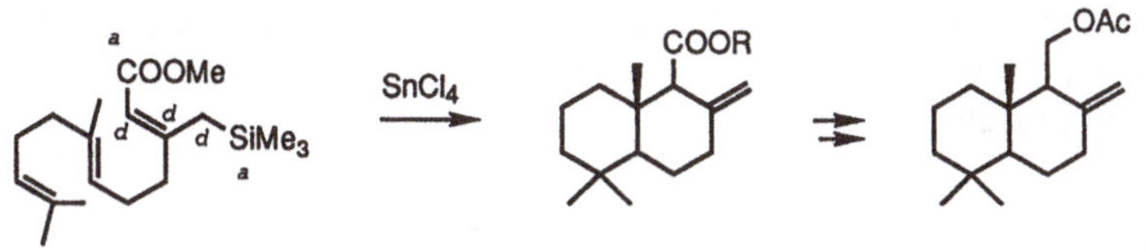

(104)

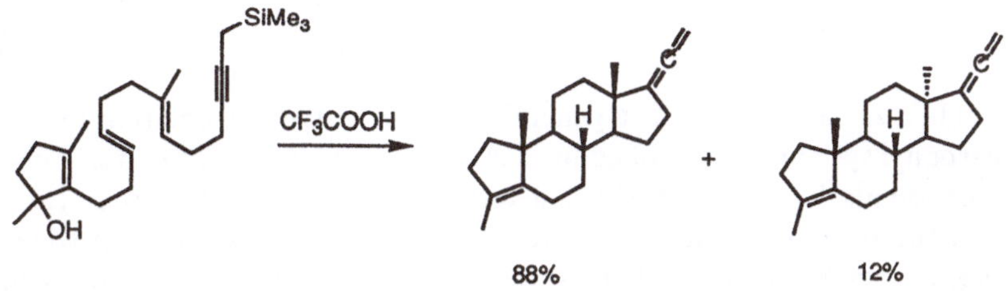

88%   12%

(105)

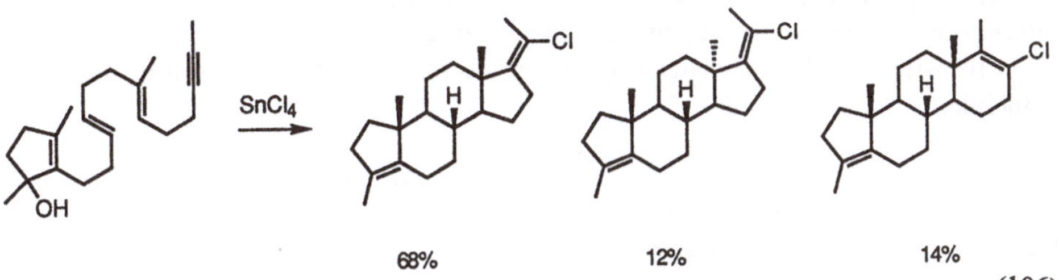

68%   12%   14%

(106)

113

Anti-Markovnikov hydration of a vinyl chloride via oxymertcuration with mercuric trifluoroacetate in methanol was surprising [128]. However, it might be speculated that solvent modified Hg(II) species show a higher affinity for the chlorine atom than the π-bond. The resulting chloronium ion is susceptible to attack by methanol in a manner analogous to Michael addition. Ejection of ClHgY is expected to be followed by a 1,2-hydride shift.

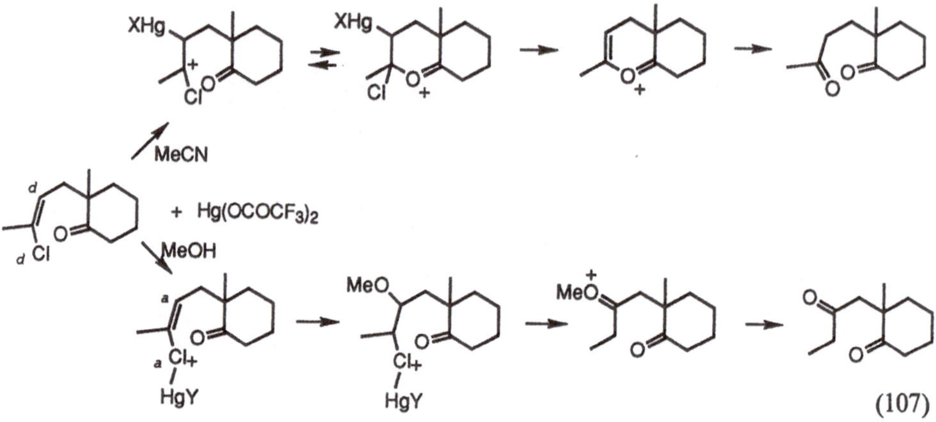

(107)

Mercury(II)-catalyzed insertion of a nitrile into the cyclobutane ring of β-pinene [129] is regio- and stereospecific. The strained C-C bond is ideally substituted such that complexation of the double bond with Hg^{2+} triggers its heterolysis to generate a relatively stable acceptor (C^+).

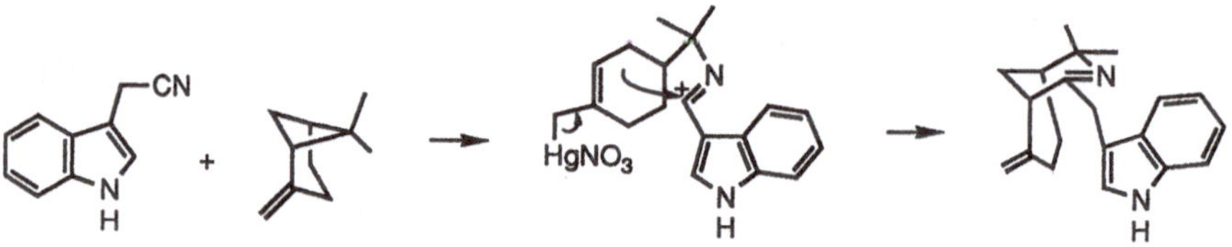

(108)

The Michael addition is a general name for conjugate addition to α,β-unsaturated carbonyl systems. The enol adducts are usually hydrolyzed to give saturated carbonyl compounds. The acceptor site is determined by the carbonyl group two bonds away, and the reaction is most efficient and selective (vs 1,2-addition) when a push-pull mechanism is operational. In other words, the reagent(s) should contain a hard acceptor and a soft donor component. Assistance of alkyl transfer from organocuprates by chlorotrimethylsilane [130] is an excellent example of such polarity alternation accentuation. Boron trifluoride etherate activation of the conjugate alkyl transfer with higher order organocuprate reagents [131] is based on the same principle.

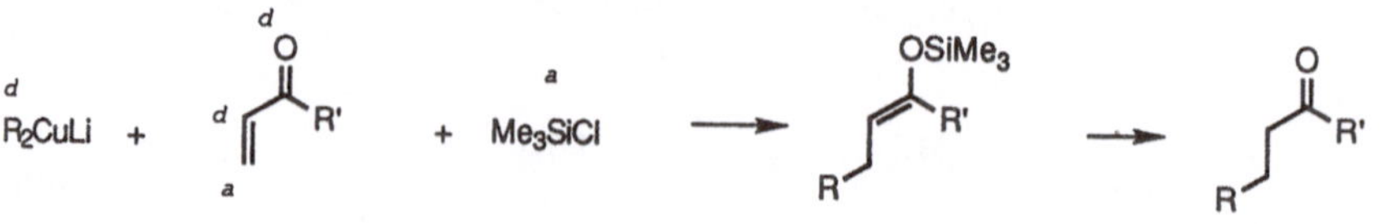

(109)

The very different reactivity toward allylation of α,β-unsaturated esters and acyl-silanes [132] is due to stabilization of the allyl cation intermediates by the alkoxy group in the former class of substances, in contrast to those derived from acylsilanes which are expected to be less stable than 1-monooxyallyl cations. In the 1,1-dioxy substituted cations two donor atoms are bonded to the acceptor site, but in the 1-oxy-1-silyl congeners a *d,a*-antagonism (captodative) exists.

$$(110)$$

Accentuation of the β-acceptor of an enone by substitution with an iminium moiety greatly contributes to the facile intramolecular Michael addition involving an enolate. The reactive components are generated from an enollactone and an ynamine [133].

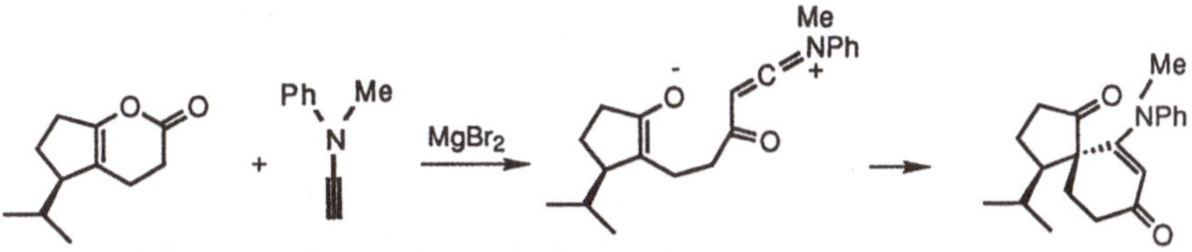

$$(111)$$

6-Amination of 5,8-quinolinedione [134] is the result of regiospecific activation through chelation.

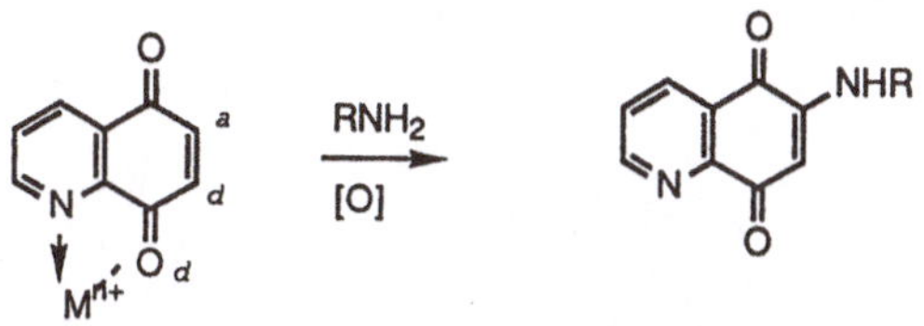

$$(112)$$

Closure of the nonaromatic ring of the anthracycline system has been effected by condensation of a 2-(3-oxobutyl)anthraquinone with nitromethane [135]. After the initial Henry reaction, a Michael-type reaction is induced by the quinone carbonyl in spite of the high electron density of the aromatic ring to be attacked, and the site of attack being a donor.

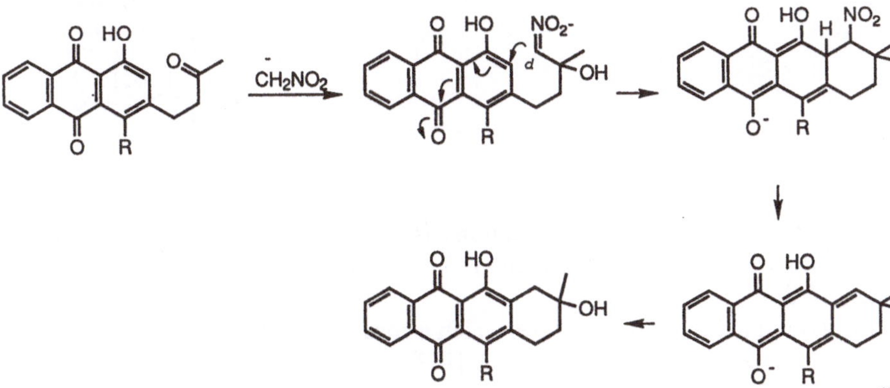

$$(113)$$

β-Alkylthio-α,β-enones undergo cyclodimerization in the presence of methyl benzoate [136]. A Claisen condensation activates the enone as Michael donor, and after cyclization by an intramolecular Michael process the benzoyl goup is eliminated.

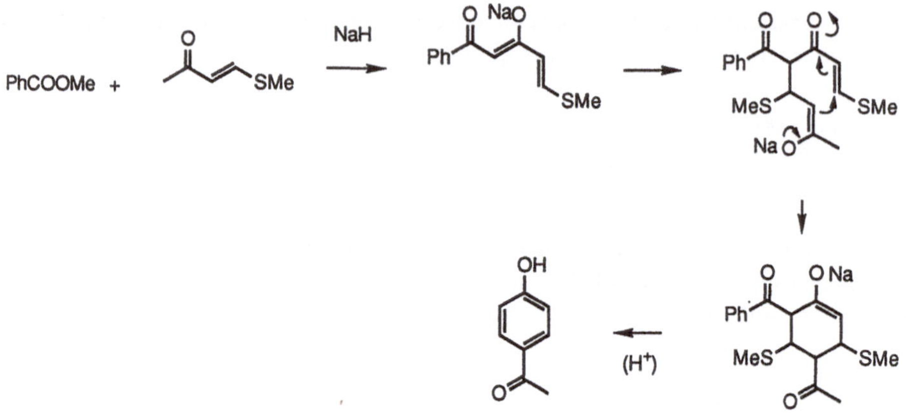

(114)

An intriguing tripartite interconversion of mitomycin A, albomitomycin A, and isomitomycin A is the result of a Michael and a retro Michael reactions [137]. The transformation of mitomycin A to albomitomycin A formally involves a *cis* addition, that from *iso*mitomycin A, a *trans* addition.

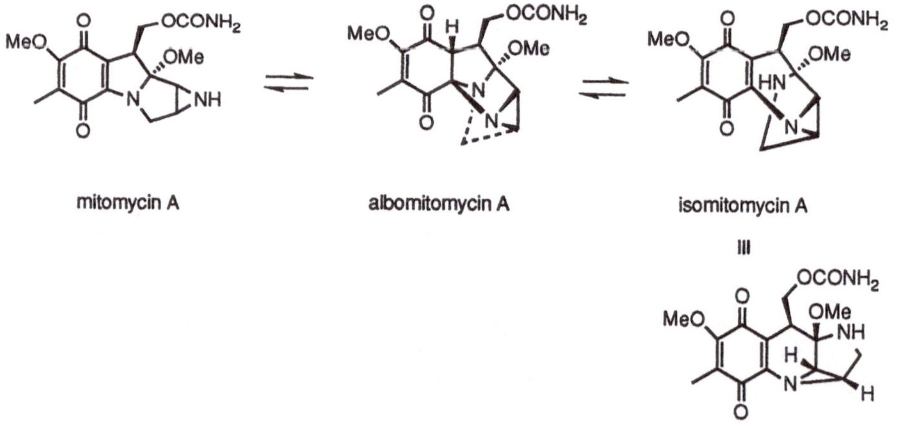

mitomycin A   albomitomycin A   isomitomycin A

III

(115)

The tandem Michael-Mannich reactions serve to assemble a tetracyclic intermediate with the aspidosperma alkaloid skeleton from 3-indoleacetyl chloride in one step [138].

The Co(I) reduction of alkyl iodides which afford nucleophilic radicals has proven useful in a synthesis of acromelic acid A [139]. In a sense contrapolarization is involved and a Michael addition follows.

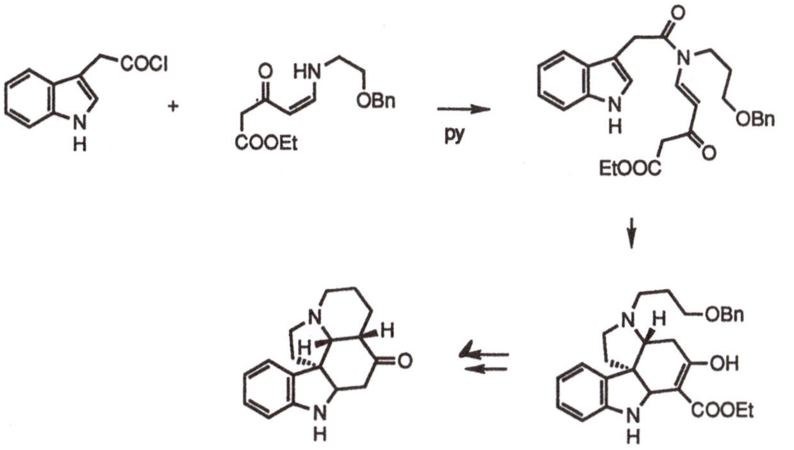

(116)

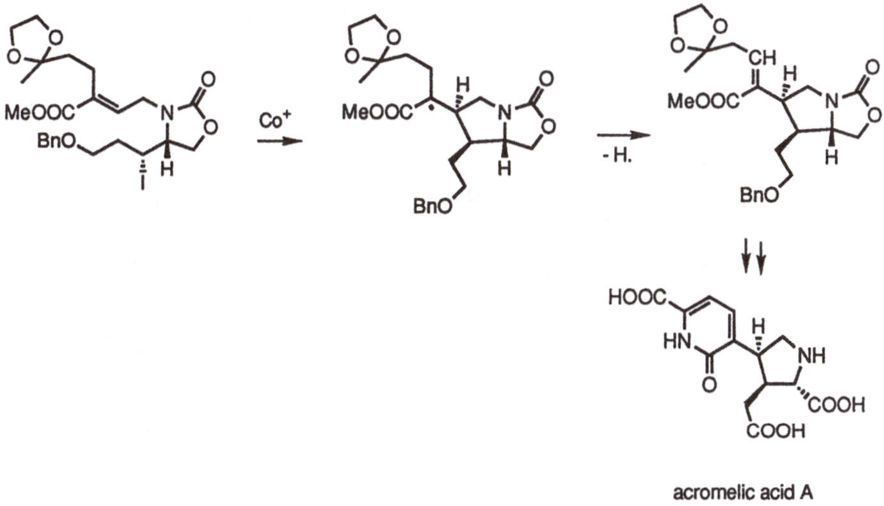

acromelic acid A

(117)

A Mannich-Michael combination is the process by which the intricate bridged skeleton of karachine [140] was constructed. One of the keys to the success is the reversibility of the Michael addition.

Very special amphiphilic properties of a divalent sulfur atom is witnessed in the reaction of α-thiomethylenecycloalkanones with acrylic esters under basic conditions [141]. The donor character of the allylic methylene group is enhanced by the sulfur (acting as an acceptor), yet the Michael cyclization that follows would be facilitated by a donor at that position.

Dithioacetalization can be regarded as a combination of addition and elimination. A cogent example of electronic control by a remote substituent is the selective reaction of 1,1,1-trifluoroalkane-2,4-diones at the 4-oxo site [142]. Here, the three

117

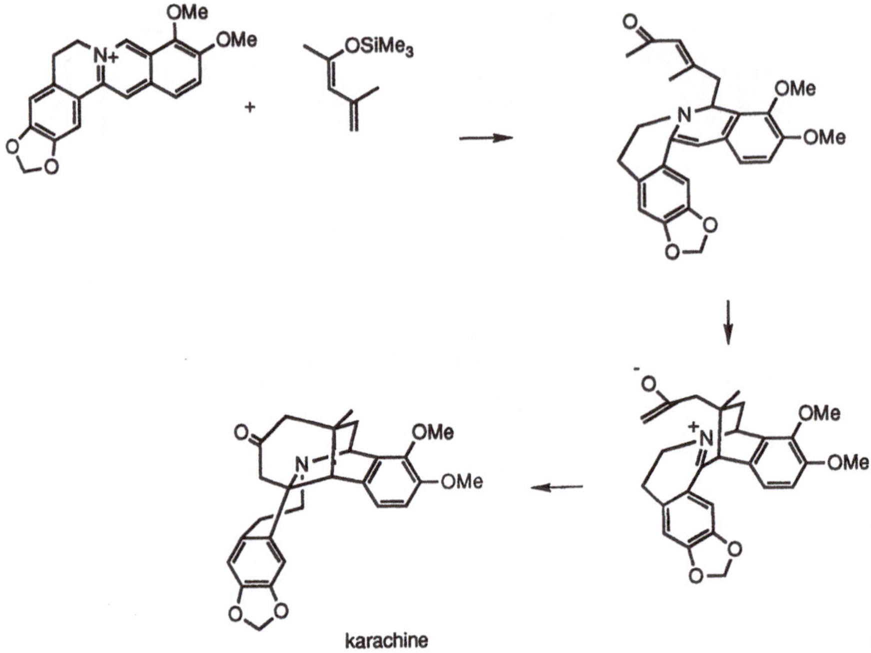

karachine (118)

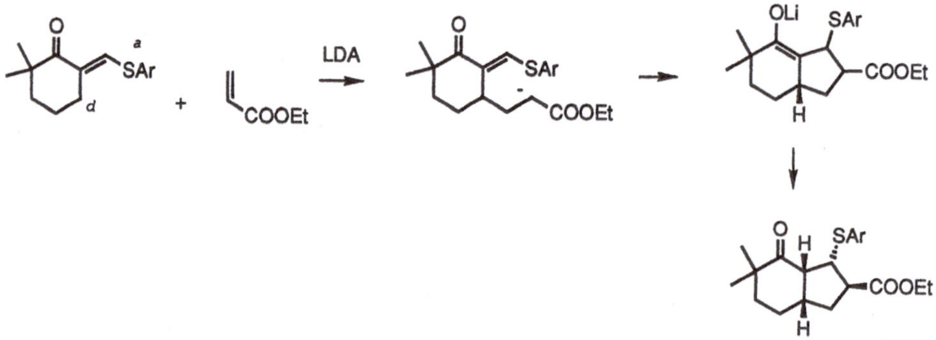

(119)

fluorine atoms imbue much donor character to the nearby carbonyl and hence a reduced reactivity toward thiols.

(120)

The high reactivity of 3-nitrophthalic anhydride (at C-1 carbonyl) towards amines [143] may be ascribed to the influence of the nitro group through polarity alternation.

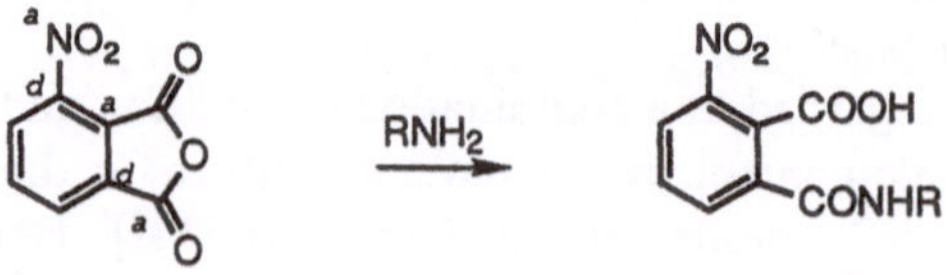

(121)

Addition to carbonyl group by a nucleophile is commonplace. However, α-lactones undergo alcoholysis with formation of α-alkoxy carboxylic acids [144]. The acceptor role of the carbonyl is restored in the bistrifluoromethyl-α-lactone [145]. It may also be considered that the α-carbon is reluctant to become an acceptor because of the fluorine atoms.

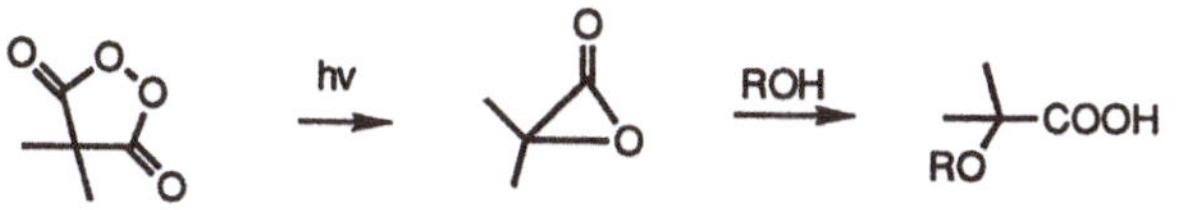

(122)

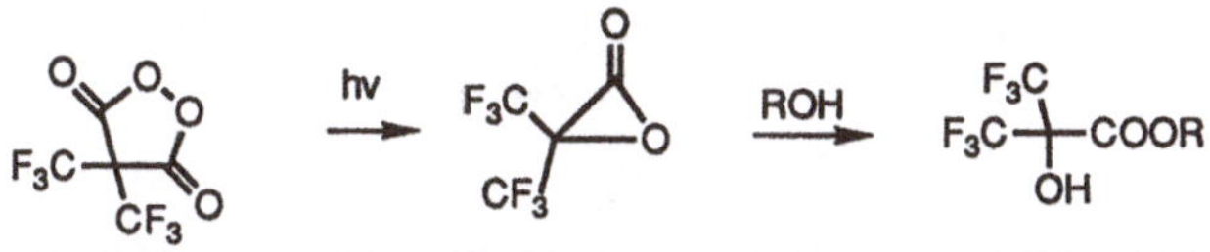

(123)

An unusual step in a synthesis of (+)-muscopyridine [146] is the methyl addition on an enolate. It appears that the oxygenated carbon still retains sufficient acceptor character by virtue of its relationship to the pyridine nitrogen (cf. reactivity of 2-vinyl-pyridine).

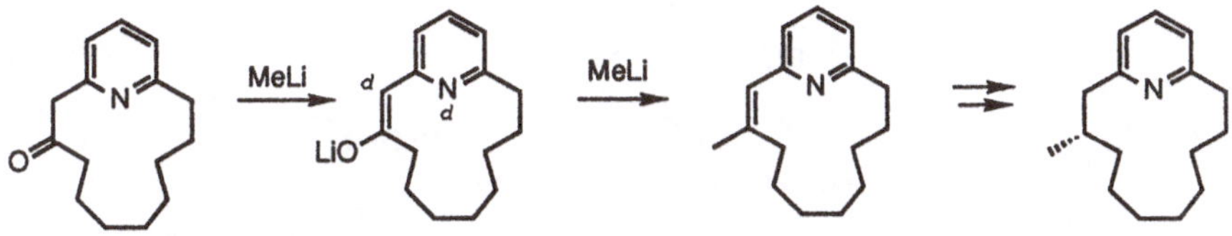

(124)

It is difficult to synthesize 9-[1-(2,4,6-cycloheptatrienyl)]-9-xanthydrol [147] from xanthone by a simple addition, considering the potential nucleophile, cyclohepta-trienide anion, is antiaromatic. An umpolung strategy works well in that reversed roles are given to the two reactants.

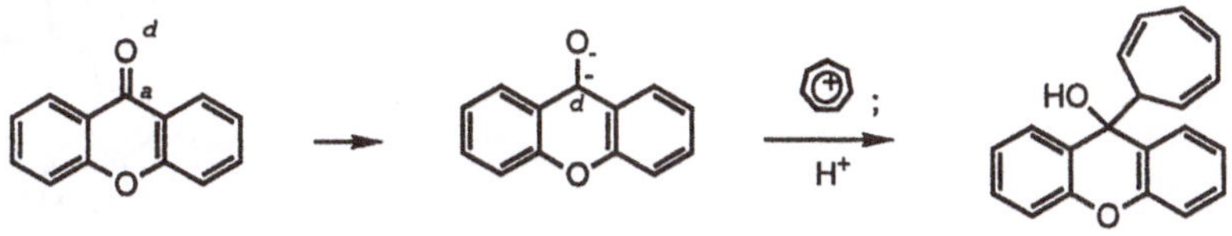

(125)

In an aldol condensation an enolate adds to a carbonyl group. Among the various methods which have been developed to achieve regioselectivity those involving polarity alternation accentuation are a time-honored technique. For example, the closure of a more strained diquinane nucleus rather than a hydrindane system is possible by installing a phosphoryl group in the α-position of the appropriate ketone [148].

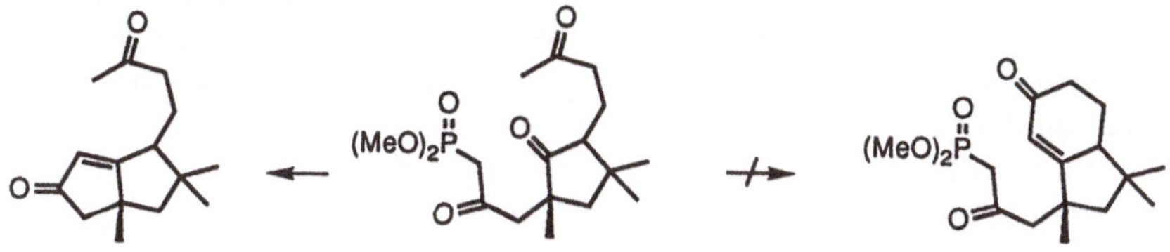

(126)

119

Interestingly, suppression of enolization of a particular carbonyl of a diketone chain is witnessed during the cyclization of 8-trimethylsilyl-2,6-octanedione [149]. Thus, enolization of the internal ketone toward C-7 is discouraged by the acceptor silicon, the competing influences of the two functions tend to diminish the donor properties of C-7. It should be noted that enolization toward C-5 leads to a nonproductive enolate, as cyclization would result in a cyclobutanol, and a fast reversal to the acyclic tautomer is expected.

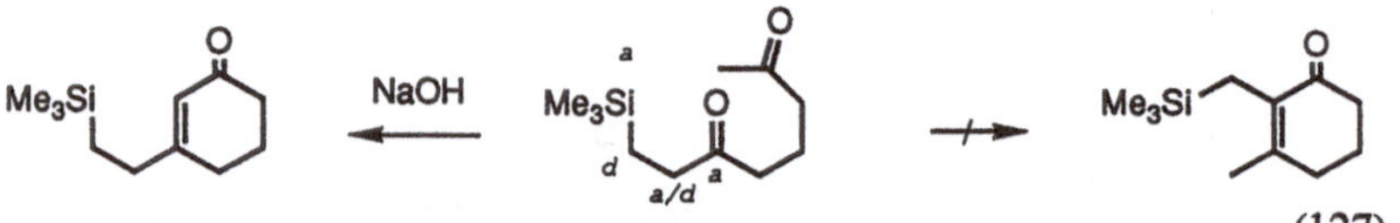

(127)

Captodative deterrence by an α-donor to guide the enolization of ketones has found an excellent application in an approach to hitachimycin [150]. Thus, aldolization of an α-methoxycyclopentanone serves to assemble the major segment of the skeletal elements.

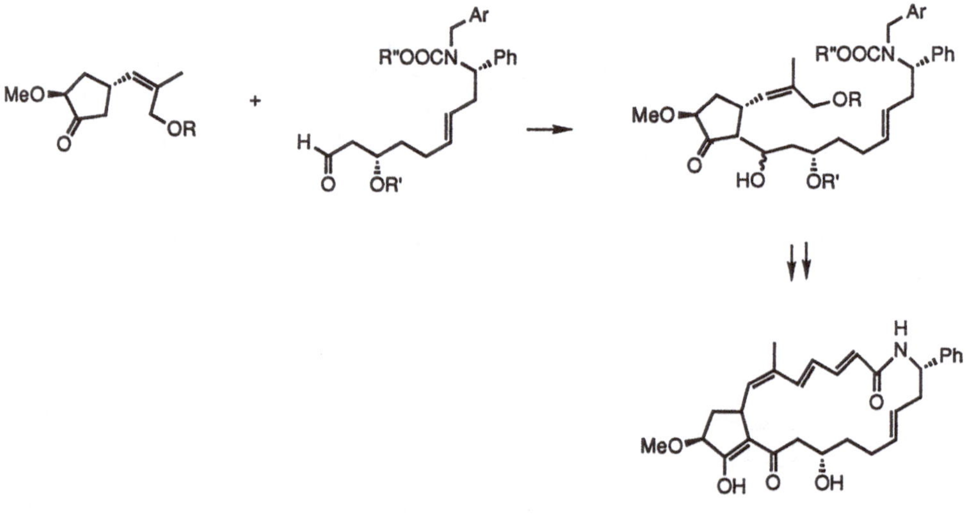

hitachimycin   (128)

Systematic bond disconnection of porantherine [151] with recognition of the double bond-carbonyl equivalence for synthesis generated a synthetic pathway which is based on two intramolecular Mannich reactions. The symmetrical nature of the amino diketone precursor identified by the retrosynthetic analysis facilitates its preparation and subsequent transformations. Moreover, all the hetero atoms (donors) are separated by odd-numbered carbon chains and such arrangements are most amenable to normal modes of assembly.

A preparation of yohimbenone [152] by closure of the D-ring with formaldehyde can be considered as a vinylogous Mannich reaction.

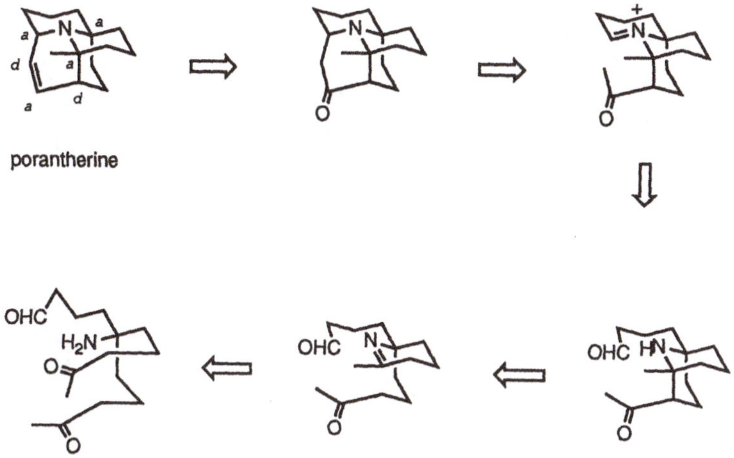

porantherine

(129)

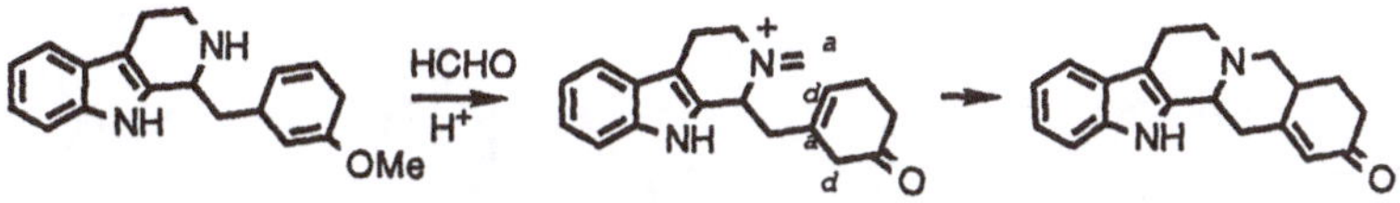

(130)

Transannular Mannich reaction of an iminium enol intermediate derived from an aza-Cope rearrangement allows access to an angular arylindolone derivative which is a synthetic precursor of crinine [153].

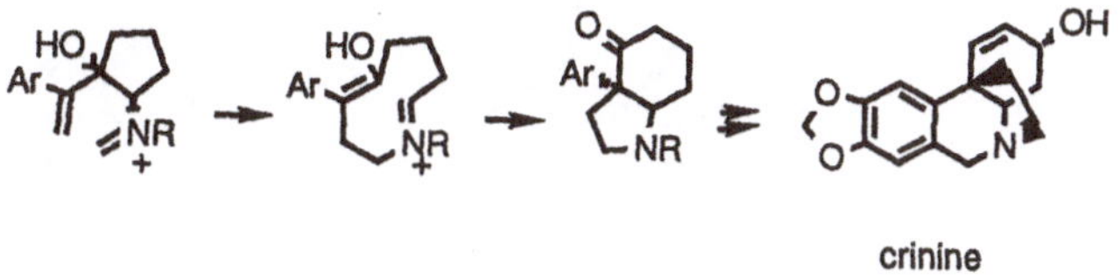

(131)

crinine

N-Acyliminium ions are versatile intermediates for synthesis of nitrogenous compounds, particularly alkaloids [154]. The conjugate system is very electrophilic such that it can be intercepted by various donors including carbonyl compounds and π-systems. In comparison with α,β-unsaturated ketones the replacement of the α-carbon with a nitrogen atom accentuates the reactivity of these species. Ingenious applications of the N-acyliminium ions include service to synthesis of corydaline [155], lycoramine [156], quebrachamine [157], and ajmaline [158], to name a very few.

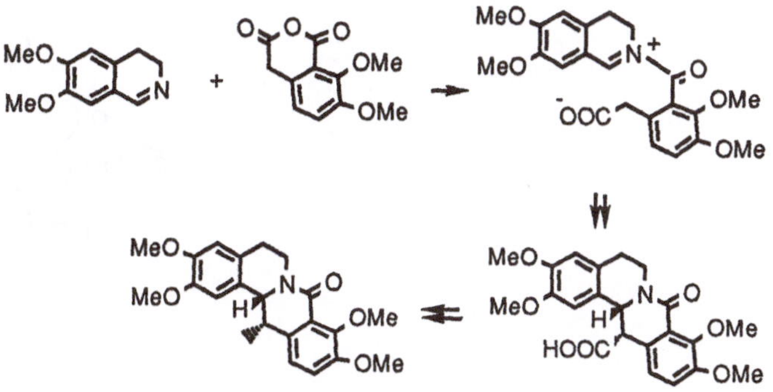

(132)

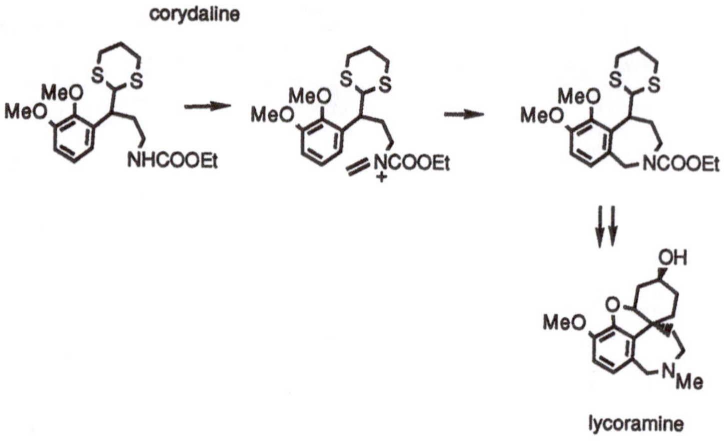
corydaline
MeO
MeO
S
S
NHCOOEt
NCOOEt
OH
O
N
Me
lycoramine

(133)

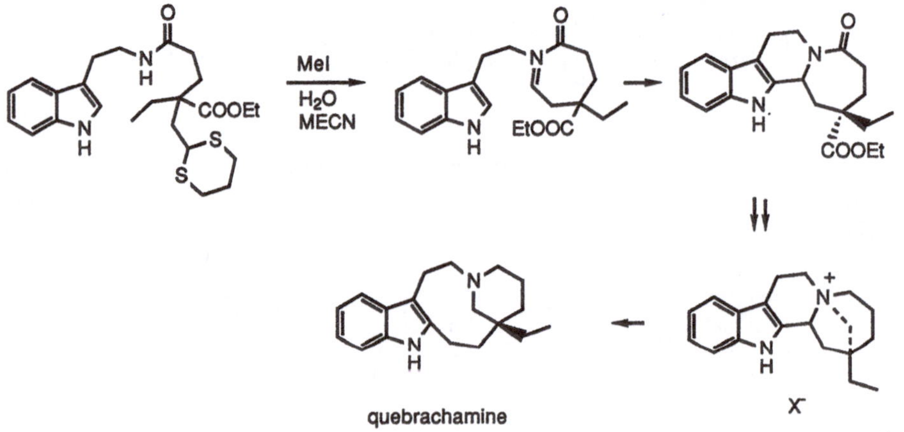
O
N
H
COOEt
S
MeI
H_2O
MECN
EtOOC
COOEt
N
X^-
quebrachamine

(134)

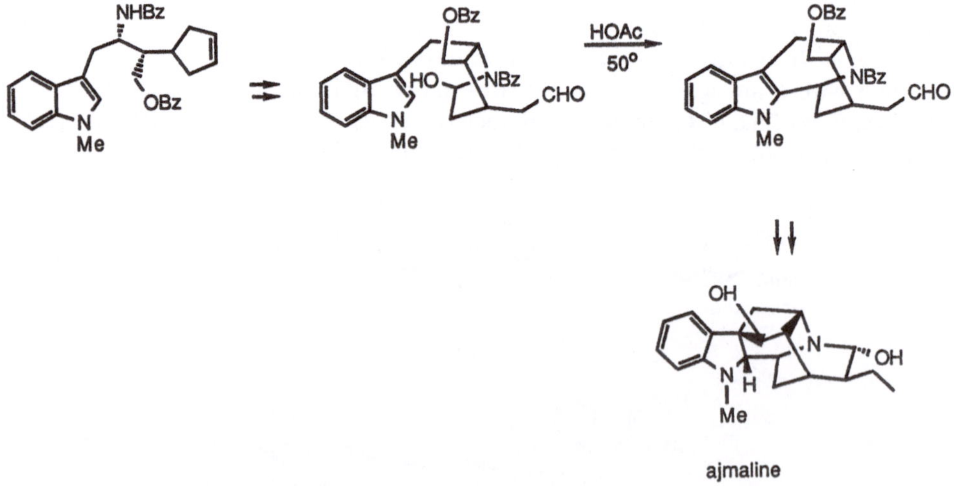
NHBz
OBz
N
Me
HO
NBz
CHO
HOAc
50°
OH
H
ajmaline

(135)

One can view *N*-acyliminium ions as activated Mannich intermediates. It is logical to expect other donors such as sulfonyl [159] can be used in lieu of the acyl moiety.

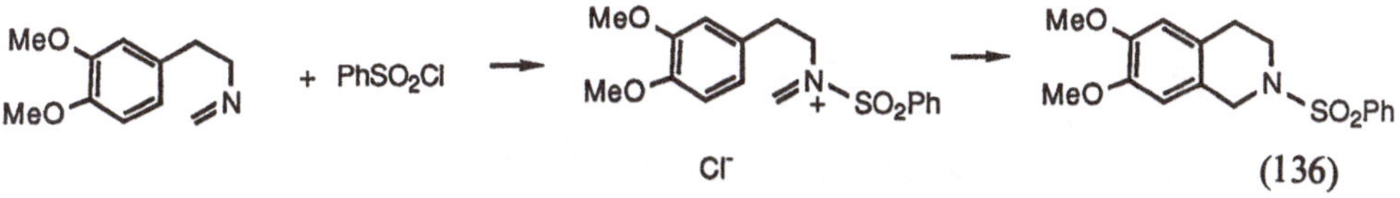

(136)

Sulfenes derived from α-halomethanesulfonyl chlorides undergo reactions with quinuclidine (used as dehydrochlorinating agent) in markedly different manners as the halogen changes from bromine to fluorine [160]. The ability of a bromine atom to assume an acceptor role which is coerced by the sulfur atom (*a-d-a* sequence) is responsible for the formation of a "dimer". On the other hand, a fluorine atom imposes its neighbor the acceptor role.

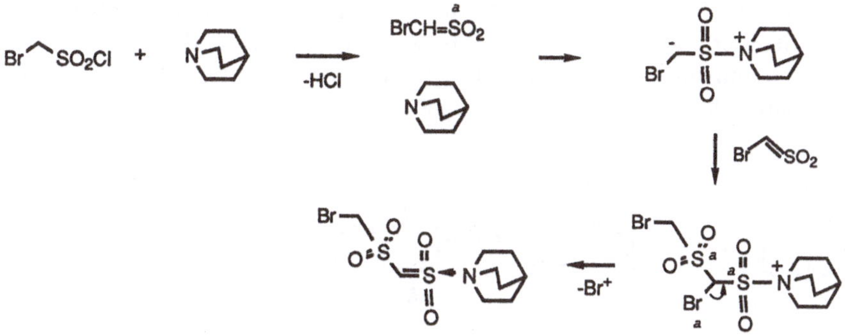

(137)

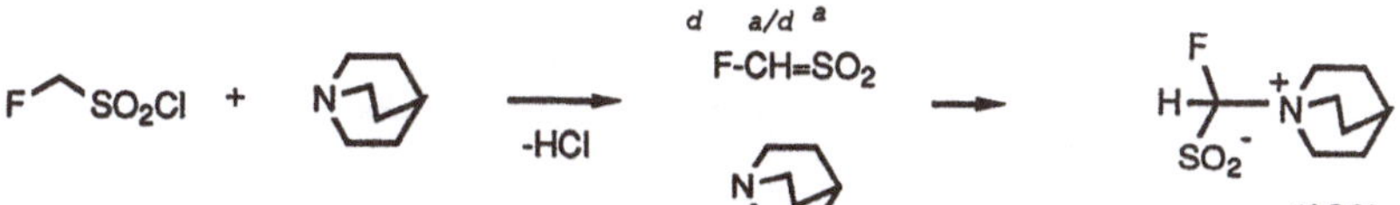

(138)

Dichlorosulfene can also behave in the umpoled fashion [161].

$$Cl_2C=SO_2 \ + \ R_3NH^+ \ Cl^- \ \longrightarrow \ Cl_3CSO_2^- \ R_3NH^+ \tag{139}$$

A method for synthesis of anthraquinones is by reaction of cyanophthalide carbanions with benzynes [162]. It is particularly useful for the access of 2-aza-1,3,8-trimethoxyanthraquinone because of the high regioselectivity imposed by the methoxy groups and the nitrogen atom of the pyridine.

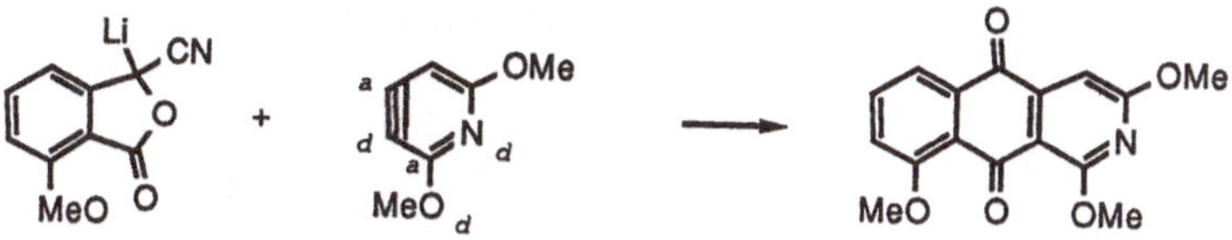

(140)

Intramolecular addition to benzyne generated from the Schiff base of aniline and *o*-chlorobenzaldehyde is thought to involve addition of an amide anion to the C=N bond [163]. This mechanism indicates donor accentuation of the *ortho* carbon.

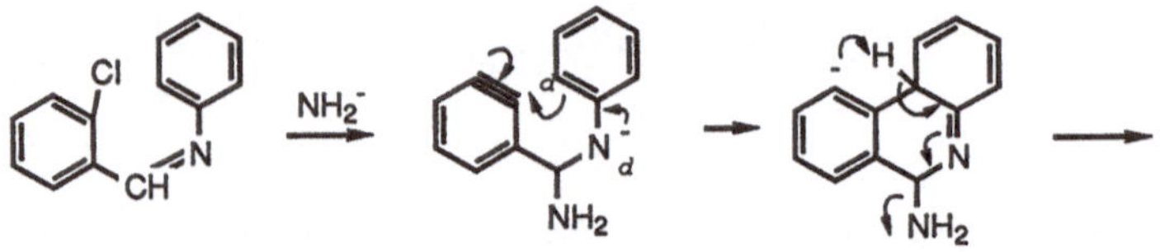

(141)

4.2 Cycloadditions

Cycloadditions give better efficiency and stereocontrol for ring formation. A high regioselectivity (95:5) for the cycloaddition of dichloroketene to a cyclohexene en route to synthetic eriolanin [164] has been observed. It appears that the allylic methoxy group play a dominant role. Note that the proximal sp^2-carbon is a donor by virtue of its 1,3-relationship with the oxygen function.

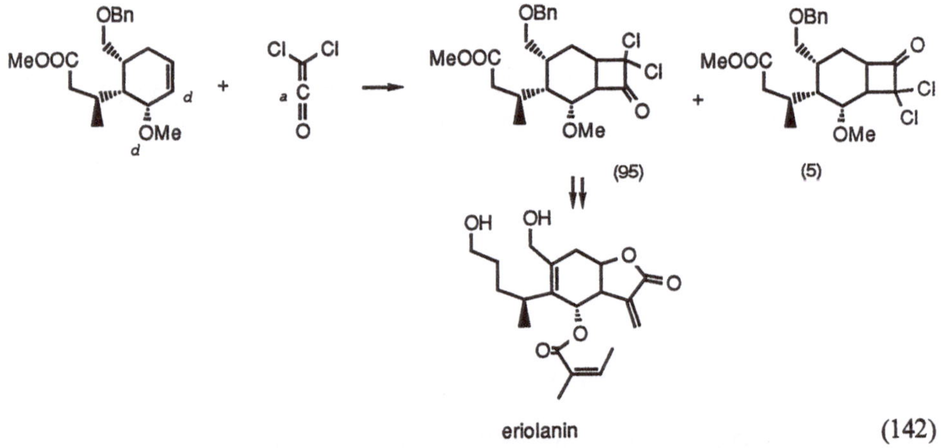

eriolanin

(142)

Dichloroketene acts as a donor toward aromatic aldehyde [165], but chlorocyano-ketene behaves as an acceptor, as shown by the reactivity profile with a series of sub-stituted benzaldehydes [97]. It must be remembered that chloro and cyano groups belong to different polarity categories, and although the fundamental donor/acceptor characters of the ketene unit do not change, the higher electrophilicity of the cyano-ketene reflects the acceptor influence by the cyano function.

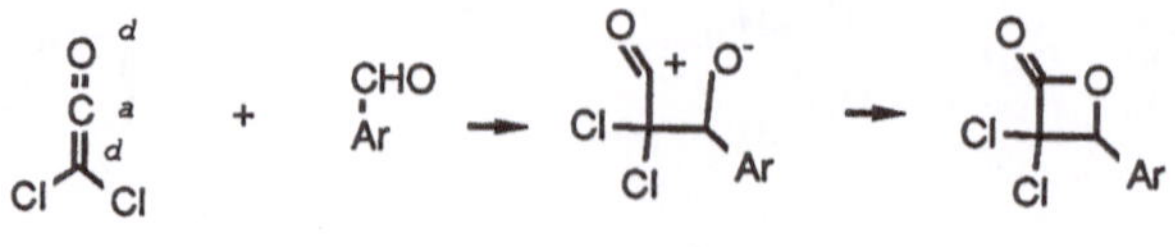

(143)

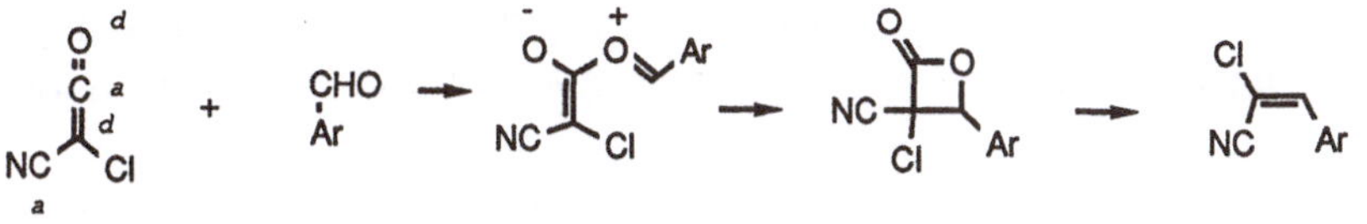

$$(144)$$

An approach to biotin [166] starts from a [2 + 2] cycloaddition of chlorosulfonyl isocyanate to chromene. The excellent regioselectivity of this reaction may be attributed, at least in part, to the ethereal oxygen.

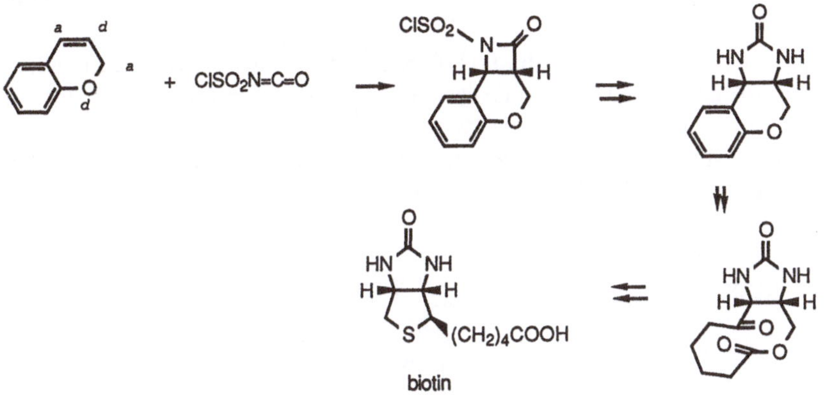

$$(145)$$

biotin

An application of the ynamine cycloaddition is found in a synthesis of dihydro-antirhine [167]. The transition state with matching polarized addends is adopted. The cycloadduct becomes fragmentable upon hydrolysis.

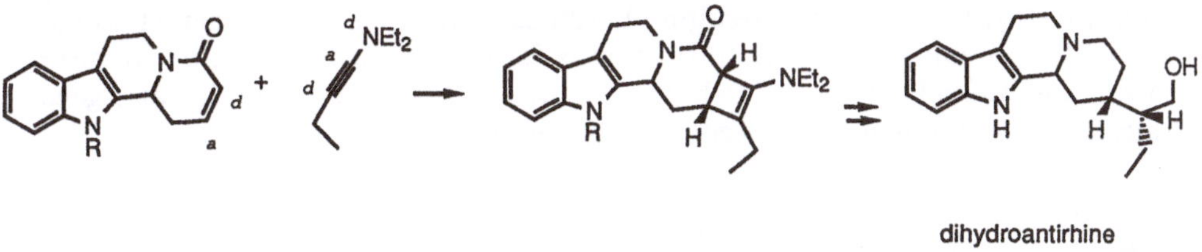

dihydroantirhine

$$(146)$$

Cycloaddition of benzo-oxanorbonadiene with chlorocyanoacetylene gives two adducts [168]. The disjointment in the cationic portion of the dipolar intermediate caused a 1,2-shift competively with ring closure.

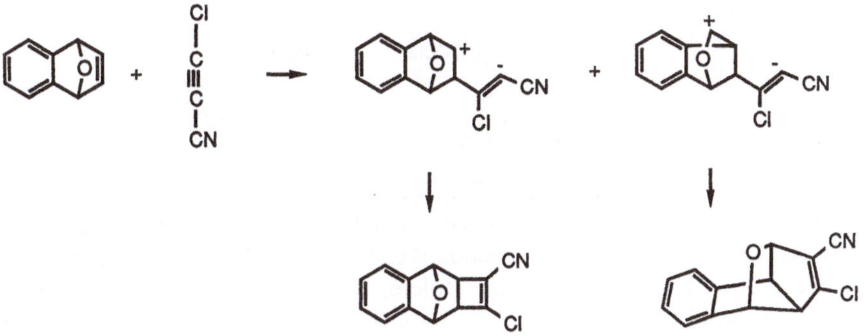

$$(147)$$

125

1-Methylpyridinium 3-oxide and benzyne form a 1:2 adduct [169]. Interestingly, in the second stage an S_N2' process is contrapolarizing with respect to the enone system.

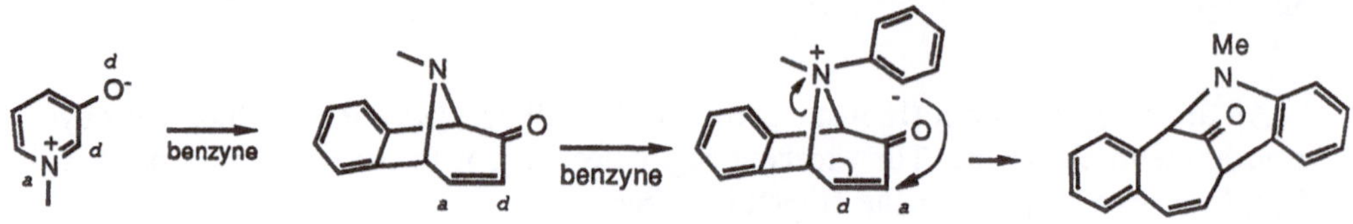

(148)

In an approach to the clavine alkaloids [170] based on intramolecular nitrone-alkene cycloaddition, the condensed tetracyclic system is probably favored by steric factors. Equilibration at higher temperature leads to predominantly a bridged oxabicyclo-[3.2.1]nonae derivative in which a portion of the molecule contains polarity alternation sequence (O–C–C–C=O, N–C–C–C=O).

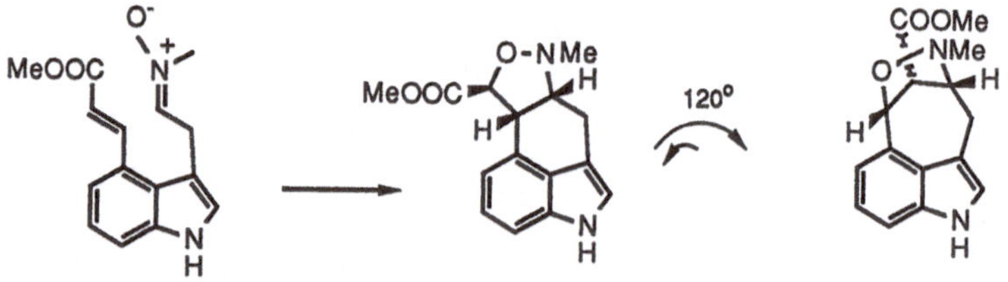

(149)

2,2-Bis(trifluoromethyl)-5(2H)-oxazolone undergoes decarboxylation on treatment with trimethyl phosphite. The resulting 1,3-dipole may be trapped by methyl propiolate [171]. The initial P–C bond formation is favored because the alternative mode of reaction would form a weak P–N bond. This counter-Michael addition also generates an a-d-a-d system which becomes fragmentable.

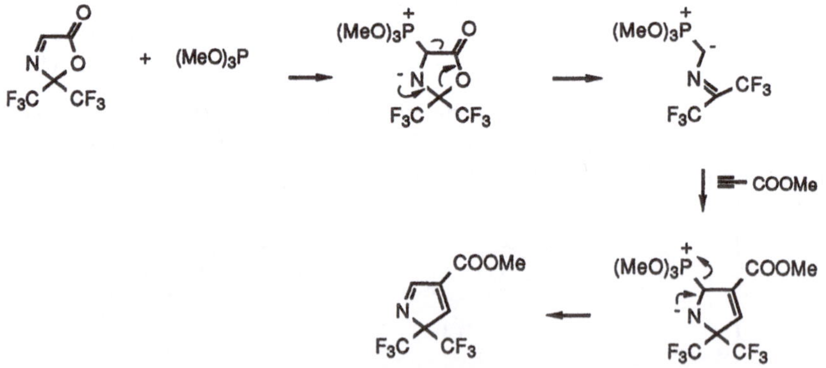

(150)

Despite the concerted nature of most Diels-Alder reactions, substituent effects are evident. Electronic compatibility of the reaction partners is of paramount importance, therefore while a normal Diels-Alder reaction is characterized by the union of an electron-rich diene and an electron-poor dienophile, the Diels-Alder reaction with inverse electron demand features an electron-poor diene and electron-rich dienophile.

Electron-rich/poor dienes do not generally condense with electron-rich/poor dieno-philes.

Given a typical diene, an increased reactivity is expected of the reaction with en-hancing polarity alternation of the dienophile. This is easily appreciated in view of an ionic reaction being more exothermic. Thus, acryloxyboranes [172] undergo cyclo-addition with cyclopentadiene even at −78 °C. The same principle underlies dieno-phile activation by replacing an enone with an alkoxyallyl cation [173] or allylidene-metalcarbonyl [174].

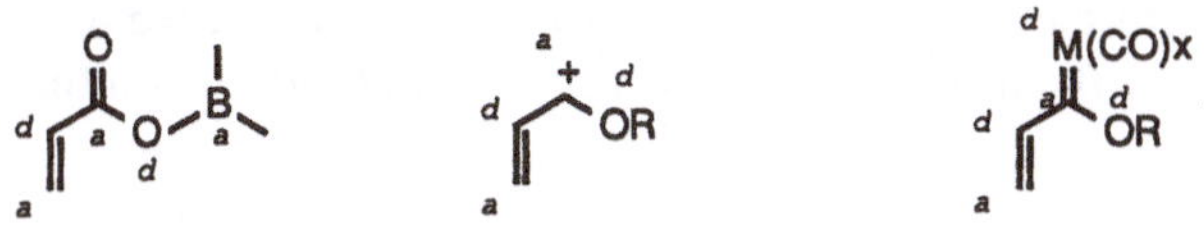

The regiochemistry of the Diels-Alder reaction of a thioaldehyde depends on the nature of the substituent directly linked to the sp^2-carbon atom [175].

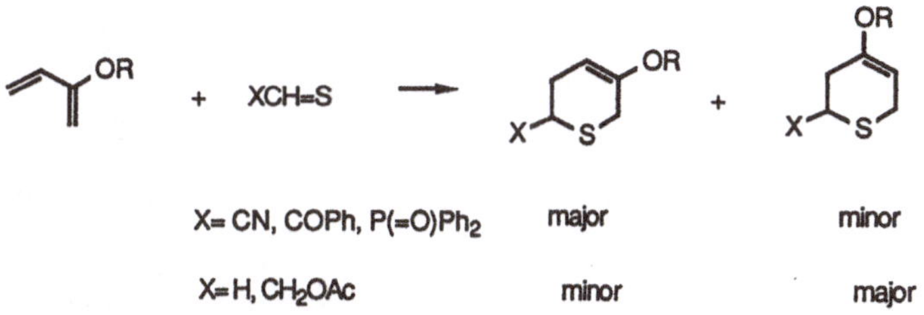

$$\text{(151)}$$

X= CN, COPh, P(=O)Ph$_2$ major   minor

X= H, CH$_2$OAc   minor   major

A peri hydrogen bonding enables juglone to partake in a Diels-Alder reaction in a regioselective fashion [176]. The H-bonded carbonyl group becomes more highly polarized which in turn causes its β-carbon more acceptor-like.

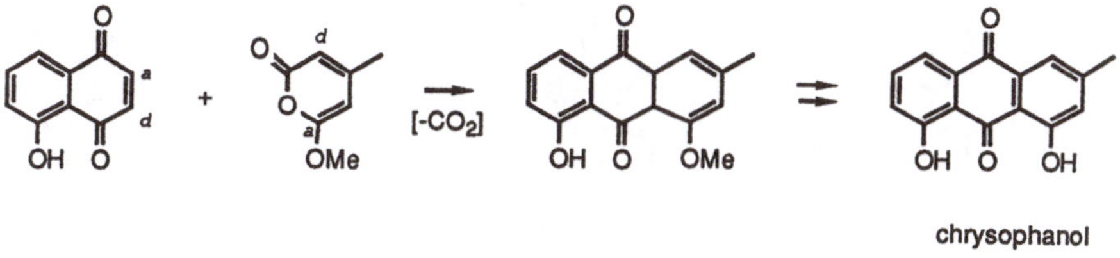

chrysophanol

$$\text{(152)}$$

Lewis acids are frequently employed in catalyzing the Diels-Alder reaction. Parti-cularly fascinating is the observation of the different regiochemistry arising from monodentate and bidentate chelation of 2-methoxy-5-methyl-1,4-benzoquinone with boron trifluoride and stannic chloride, respectively [177].

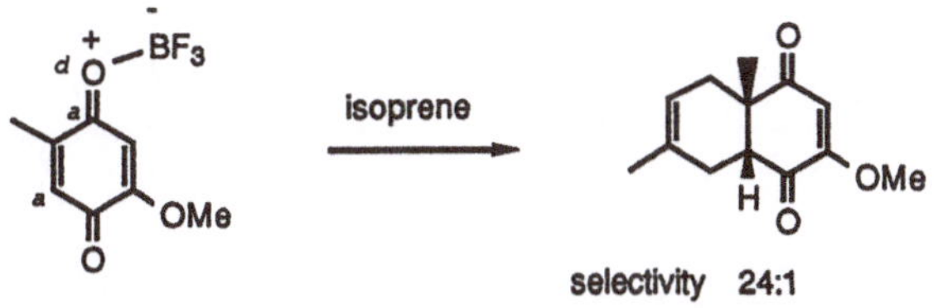

isoprene

$$\text{(153)}$$

selectivity 24:1

127

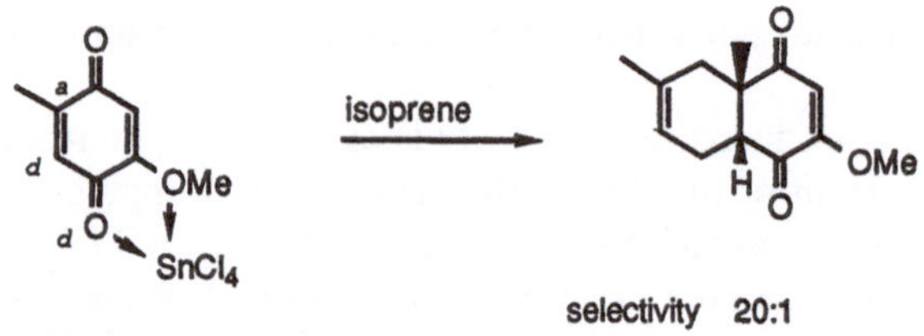

$$(154)$$

selectivity 20:1

The regioselective formation of a *cis*-hexalone containing a homonuclear diene which is useful for elaboration into occidentalol [178] by the condensation of 4-methyl-3-cyclohexenone with methyl coumalate is rather surprising. It seems that C-3 of the cyclohexenone, under the influence of the ketone group (perhaps via the enol), prefers an acceptor role.

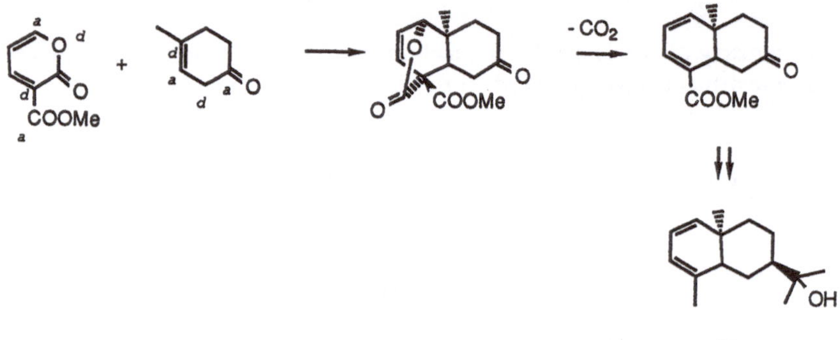

occidentalol $$(155)$$

The most efficient and regioselective Diels-Alder reactions are those involving electronically compatible dienes and dienophiles. It might be instructive to mention the rapid syntheses of olivacine [179] and catharanthine [180] benefit from the well-designed reactions.

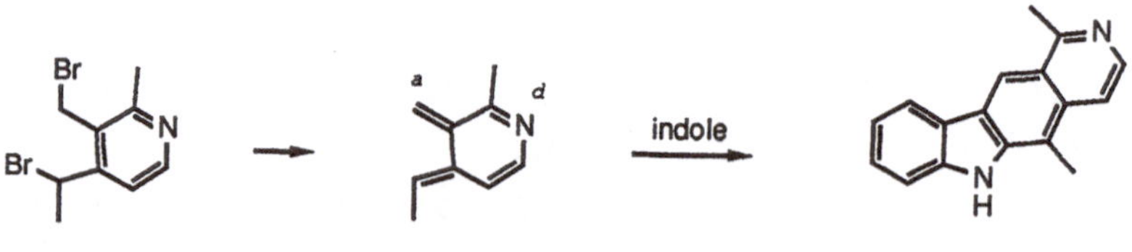

olivacine (156)

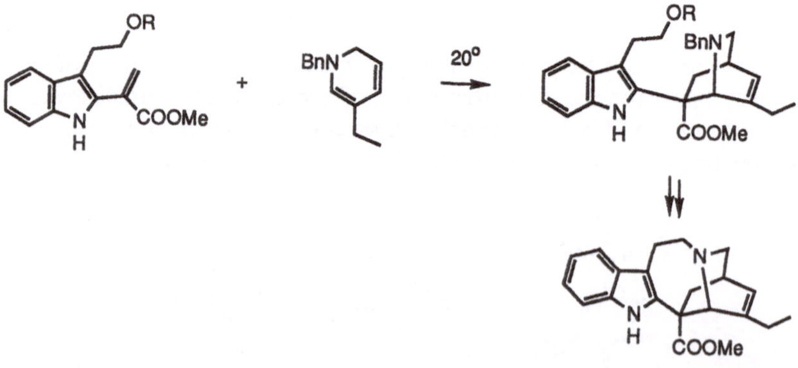

catharanthine (157)

The study of the effects of competing functionalities often contributes to better understanding of a reaction. Thus, the regiocontrol, i.e. p-directing, a sulfur substituent at C-2 of 1,3-butadiene exerts has been shown to depend on its oxidation state [181], an observation in total agreement with the gradual transition from a donor to an acceptor.

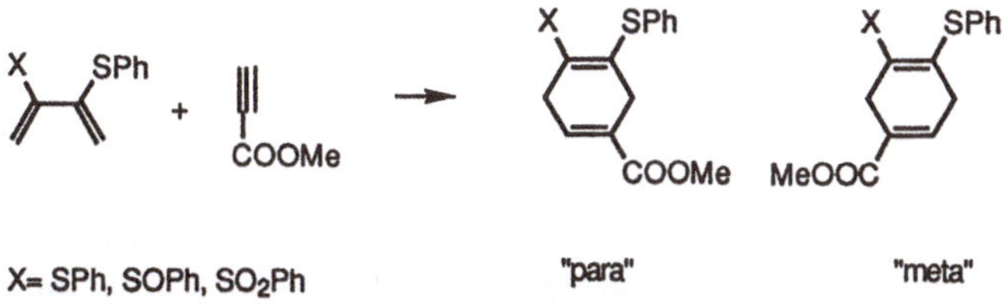

X= SPh, SOPh, SO$_2$Ph   "para"   "meta"   (158)

Complete regiodirection by the NHCOOR for 4-sulfurated (sulfenyl, sulfinyl, and sulfonyl) E-1,3-butadiene-1-carbamates in their Diels-Alder reactions [182] is quite natural. The case at issue might be the sulfenylated compounds, but we must remember that divalent sulfur can act as an acceptor.

The strong directive effect of an allylic silicon in a diene on the regiochemistry of the Diels-Alder reaction proved to be a blessing in synthesis of 11-deoxycarminomycinone [183].

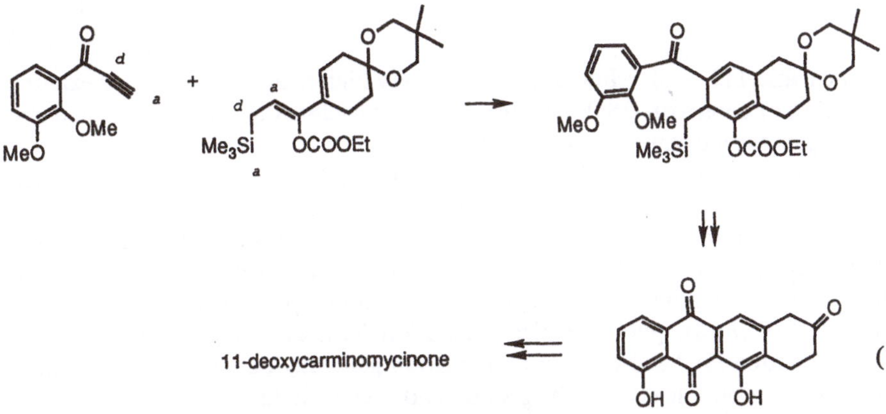

11-deoxycarminomycinone   (159)

Simple polarity alternation arguments readily accommodate the regiochemical outcome of the Diels-Alder reactions of 5,6-bis(methylene)-2-bicyclo[2.2.1]heptan-2-one [184] and a lactone analogue [185].

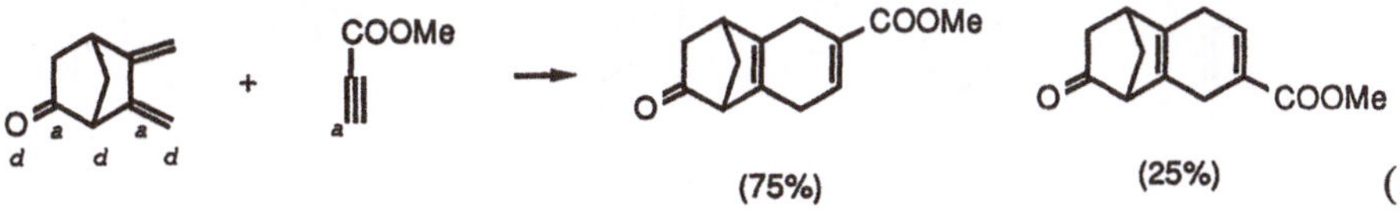

(75%)   (25%)   (160)

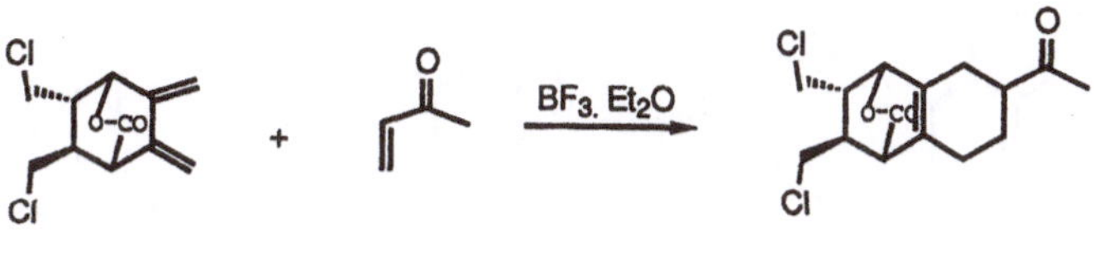

BF$_3$, Et$_2$O   (161)

129

However, in these ring systems through-space homoconjugative interactions are sterically very favorable, and the through-bond (polarity alternation) interactions may be overwhelmed [186].

$$(162)$$

Cationic [5 + 2]cycloaddition involving a styrene and p-benzoquinone ketals in the presence of a Lewis acid is the key reaction to a number of neolignans (e.g. guianin [187]. The oxygen atoms of the dioxolane function in the cycloaddend reinforce the a character of their neighboring carbons, and this character is transmitted to the termini which participate in the C-C bond formations.

guianin

$$(163)$$

Photochemical [2 + 2]cycloadditions show very high regioselectivity. For example, dimerization of coumarin [188] gives head-to-head cycloadducts (singlet excited state → syn, triplet excited state → syn + anti).

More complex systems include α-lumicolchicine [189] which is an anti/head-to-head cycloadduct of the electrocyclized β- and γ-lumicolchicine.

The cycloaddition of an enone with electron-rich alkenes also proceeds with remarkable regioselectivity. The rationalization is that in the excited state, polarization of the enone double bond is opposite in direction as compared with the ground state [190]. In other words, photochemical excitation induces contrapolarization. With this consideration the head-to-head dimerization generally observed is reasonable as it involves one molecule each in the ground and excited states.

Elegant applications of the contrapolarized orientation in such photocycloadditions include syntheses of caryophyllene/isocaryophyllene [80], 10-epijunenol [191], and helminthosporal [192].

caryophyllene

$$(164)$$

10-epijunenol

$$(165)$$

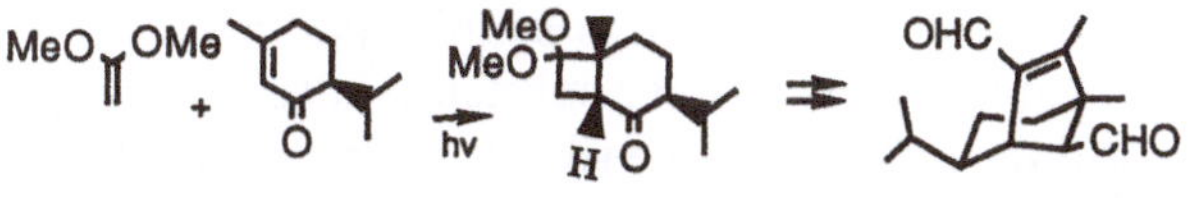

(166)

helminthosporal

It is pertinent to mention the formation of γ-ketols by photochemical addition of alcohols to enones follows the same polarity pattern [193].

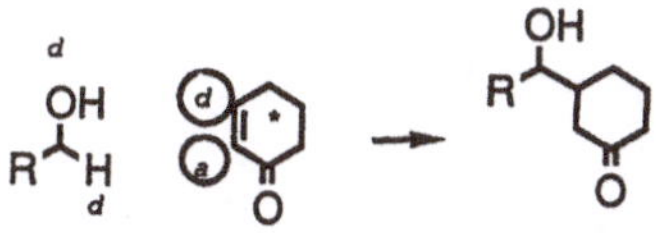

(167)

Thermally and photochemically induced cycloadditions of 2,3-butanedione with 1,1-diethoxylene show different regiochemistry [194]. This is expected on the basis of the above explanation. Perhaps electron transfer occurs between reactants before the bond formation starts.

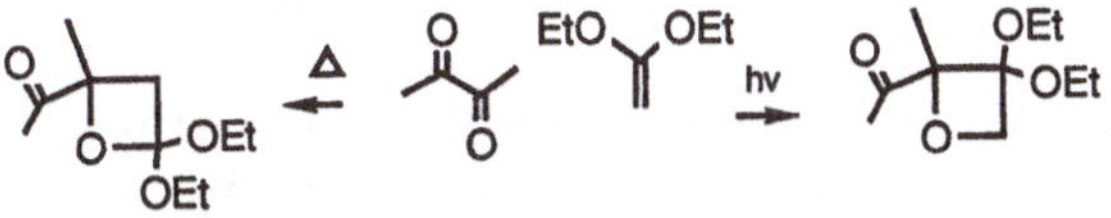

(168)

Concerning the regiocontrol of intermolecular Pauson cyclization in which a homo-allylic donor group is present in the alkene subunit [195], there seems to be some electronic contributions.

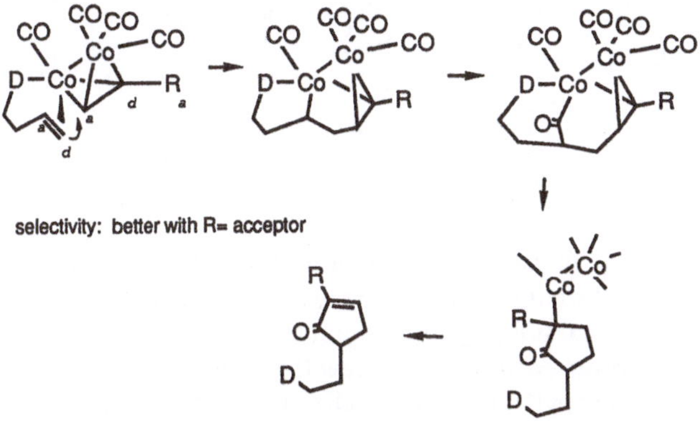

selectivity: better with R= acceptor

(169)

4.3 Electrocyclizations and Cycloreversions

The Nazarov reaction [196] is a conrotatory electrocyclization involving four electrons over a five-carbon span. Usually, a more highly substituted cyclopentenone is obtained. However, contrathermodynamic products may be generated by placing a silyl group at the β-position of a bare vinyl moiety in the cross-conjugated dienone [197]. The acceptor facilitates and controls the regiochemistry of the cyclization process.

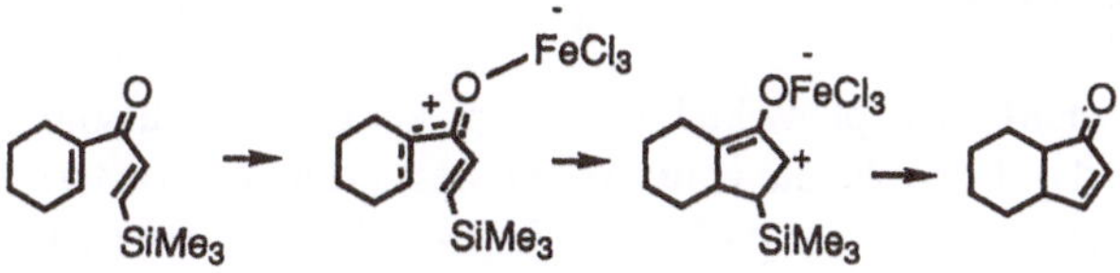

(170)

131

The rationalization for the regiocontrol is simple. Participation of the silylated double bond is disfavored on account of the substantial positive charge to be alloted to the carbon which is directly linked to the silicon atom, an *a-a* arrangement.

The intramolecular ene reaction [198] proceeds in good reioselectivity if either the ene or the enophile component is linked to polarizable groups. Based on the direction of polarization an excellent synthesis of (—)-α-kainic acid [199] has been designed.

$$\text{(171)}$$

$$\text{(172)}$$

(-)-α-kainic acid

The relative rates for electrocyclic ring openings of cyclopropyl ions [200] are shown in Eqs (173, 174). For the cations the faster rates are exhibited by compounds in which an acceptor is directly bonded to the electron deficient center (*a-a* arrangement), whereas precursors with a donor substituent at the center open most slowly.

$$\text{(173)}$$

$$\text{(174)}$$

The phenomena are explainable in terms of destabilization of the former series of cyclopropyl cations by the contiguous *a-a* array as well as stabilization of the resulting allyl cations by a beneficial polarity alternating sequence. The stabilization/destabilization situations are reversed in the least reactive compounds.

In the cases of cyclopropyl anions, destabilization by the *d-d* pairing and stabilization in the open forms for the most reactive substrates, and the opposite scenario for the least reactive substances are evident.

5 Rearrangements

5.1 1,2-Rearrangements

Most 1,2-rearrangements are provoked by the presence of a disjoint sequence of atoms. In other words, the molecule is not in the best electronic order. For example, epoxides, glycols and β-amino alcohols contain two *d-a* segments which are linked head-to-head. Proper activation such as protonation or diazotization would induce a rearrangement to remove the unfavorable electronic interations.

It must be pointed out that the 1,2-alkyl shift may be subverted by other factors such as steric strain. Thus the cascade rearrangement of a tetraspiroketone on exposure to acid is most aesthetically appealing and synthetically useful for entry into [4.5]coronane [201].

$$\qquad\qquad (175)$$

The thermal isomerization of a spirocyclic enol ether to the ketone [202] (Eq. 176) is probably a homolytic process. However, it is noted that part of the driving force for the reaction must be the bonding of the ethereal oxygen to a designated donor atom of the cross-conjugated cyclohexadienone moiety.

$$\qquad\qquad (176)$$

Intramolecular interception of an N-alkoxy-N-acylnitrenium ion by an aromatic ring results in a spirocyclic species [203]. There follows a C-migration (instead of N-migration) to restore aromaticity. This migration is assisted by the stabilization of the incipient carbocation by the resident donor nitrogen atom; N-migration would not profit from this favorable electronic interaction.

$$\qquad\qquad (177)$$

A convenient preparation of 2-deoxy sugars [204] from glycosyl halides is based on the rapid and irreversible migration of an acyloxy group from C-2 to the anomeric center upon generation of the glycosyl radicals. This is a thermodynamically controlled reaction because it restores a d-a-d arrangement.

$$\qquad\qquad (178)$$

An intriguing C-C bond scission of camphorquinone is via rearrangement [205]. The rearrangement, which is induced by a destabilizing a-a disposition in addition to ring strain, transfer an acceptor site to an adjacent atom and thereby creating a fragmentable situation.

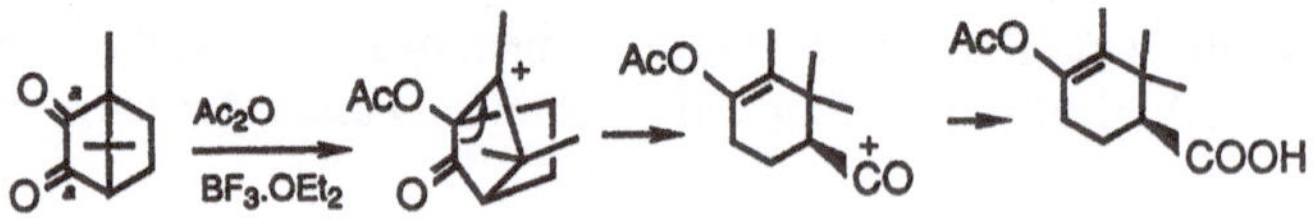

$$(179)$$

Disjoint functionalities also exist in chloral and this property has been exploited in the isomerization of 4-hydroxy-2-cyclopentenones [206].

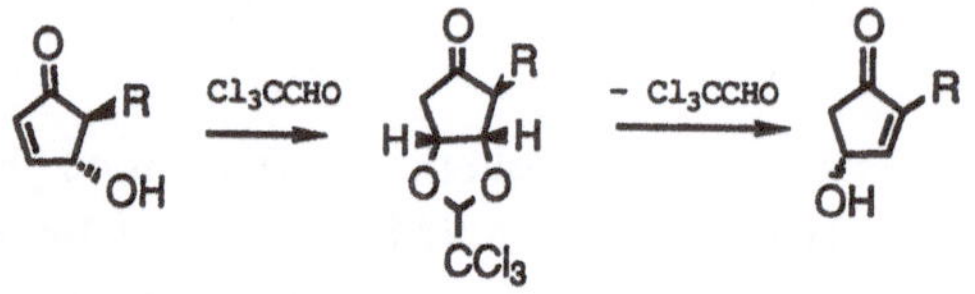

$$(180)$$

trans-2-(Trimethylsilyl)methylcyclohexyl *p*-nitrobenzoate is also disjount. Fluoride ion promotes a ring contraction [207] which also removes the undesirable polar interactions.

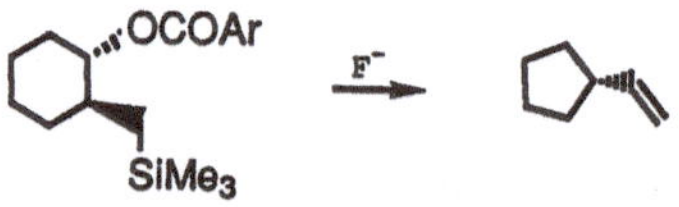

$$(181)$$

2-Phenylthio-3-trimethylsilylalkanols lose the element of trimethylsilanol on treatment with an acid [208]. Ionization of the hydroxyl group is participated by the sulfur atom and the episulfonium intermediates then undergo desilylation. The Si-C-C-S segment is conjoint and its fragmentation proceeds upon activation by episulfonium ion formation.

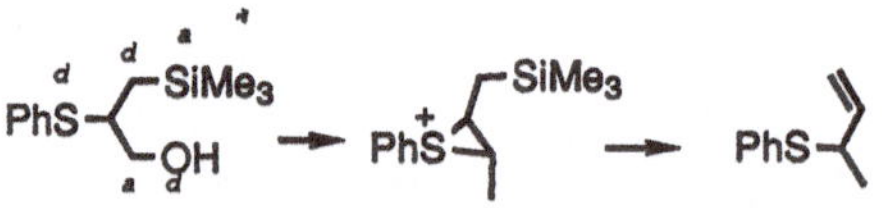

$$(182)$$

A silyl group directs ring expansion of a ketoaldehyde [209] via interaction with the incipent cation.

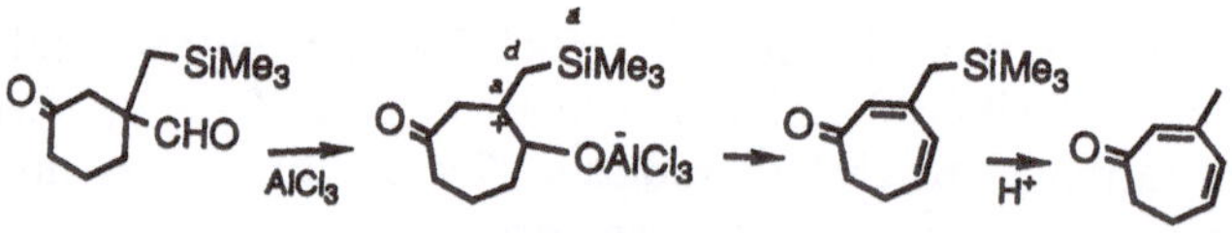

$$(183)$$

The attempted epoxidation of α-alkylidene cyclic sulfones led directly to enlarged sulfonyl cycloalkanones [210]. The rearrangement step is thermodynamically controlled in that products contain an *a-d-a* sequence.

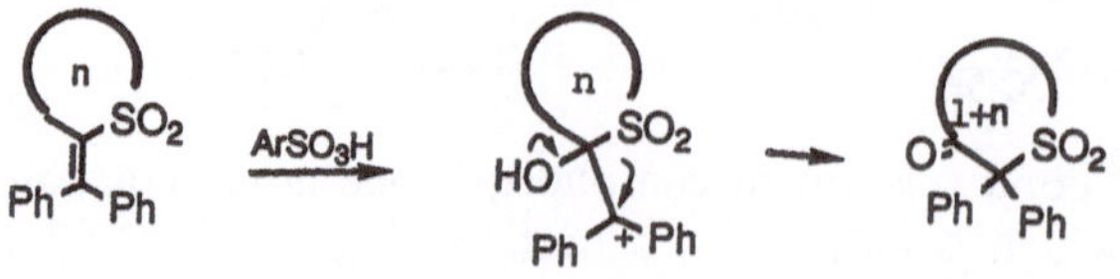

$$(184)$$

A fascinating contrapolarization-driven reaction is the conversion of a protoberine skeleton to an ochotensine-type alkaloid [211]. The quinomethide carbon in the seco intermediate assumes a *d*-character which is made possible by electron supply from the oxy substituent at the meta position. Note the C-N bond cleavage step entrusts this carbon an acceptor role.

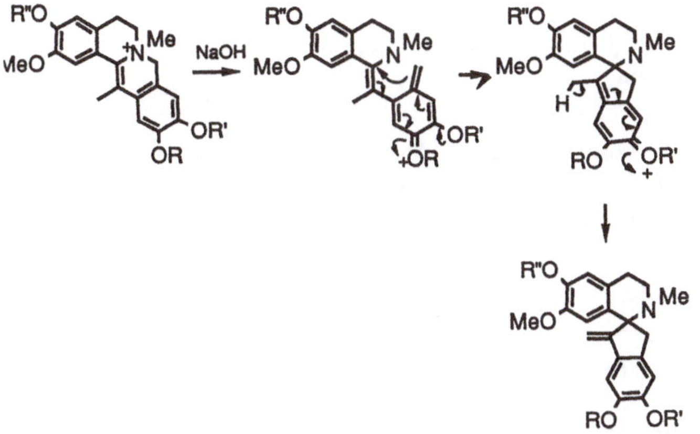

(185)

In a synthesis of aspidospermidine [212] an alkylative rearrangement was employed to gain entry into the pentacyclic skeleton.

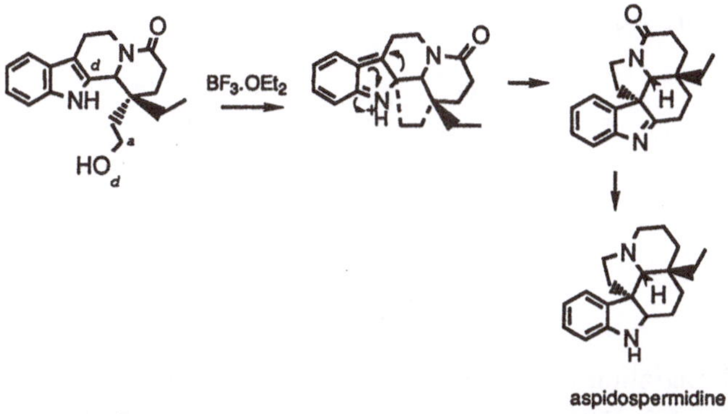

(186)

aspidospermidine

Many other 1,2-rearrangements such as Hofmann, Curtius, Beckmann rearrangements and Baeyer-Villiger reaction can be rationalized in terms of response to unfavorable *a-a* situations.

The Favorskii [213] and Ramberg-Bäcklund [214] rearrangements are mechanistically akin to the above. The first step is deprotonation (although in highly strained systems the Favorskii rearrangement proceeds by donor addition to the carbonyl group) which is followed by internal displacement of the α-halogen.

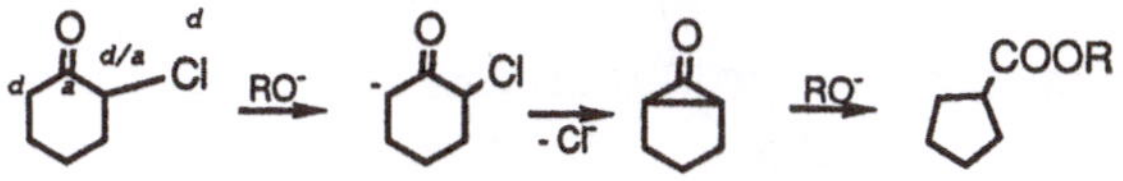

(187)

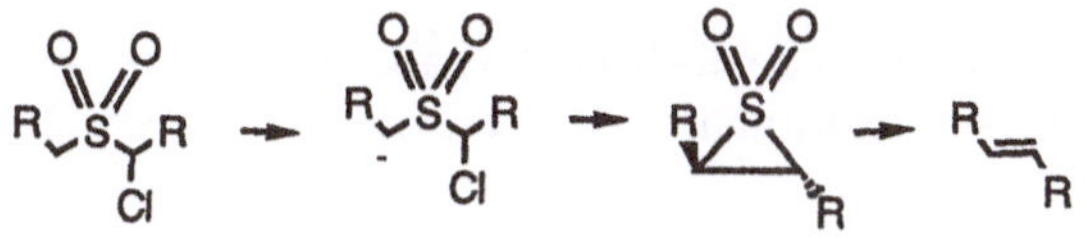

(188)

The apparent suppression of deprotonation at the α-carbon may be due its capto-dative nature. This carbon atom is bonded to both an acceptor (C=O) and a donor (Hal), the corresponding carbanion is destabilized by the donor substituent. In con-nection with this tendency is the selective silylation [215] of 1,3-oxathian-3,3-dioxide at C-4.

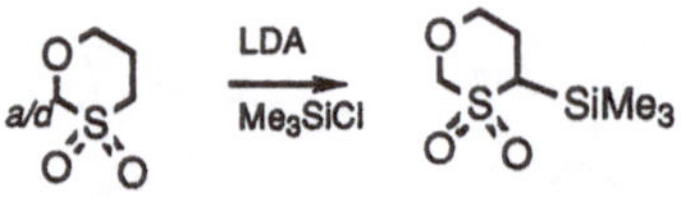

$$ \text{(189)} $$

Treatment of pulegone dibromide with sodium hydroxide leads to a cyclopentane-carboxylic acid [216]. Unidirectional opening of the cyclopropanone intermediate is caused by the bromine atom at the β-position which acts as a donor.

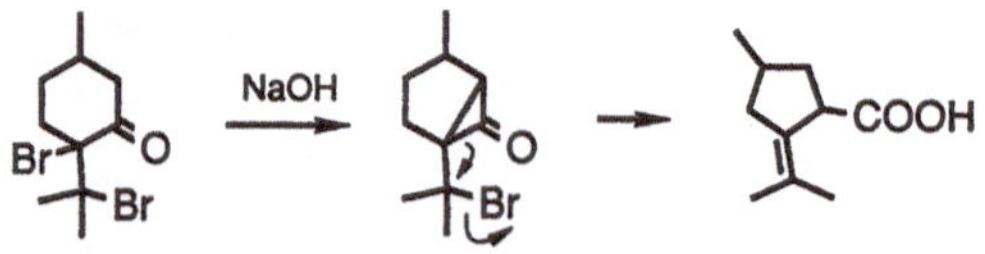

$$ \text{(190)} $$

The desulfurative C-C bond formation process developed by Eschenmoser [217] is somewhat akin to the Ramberg-Bäcklund rearrangement. An episulfide intermediate is formed, which on being attacked by a phosphine, begets a carbon skeleton with polarity alternating sequence.

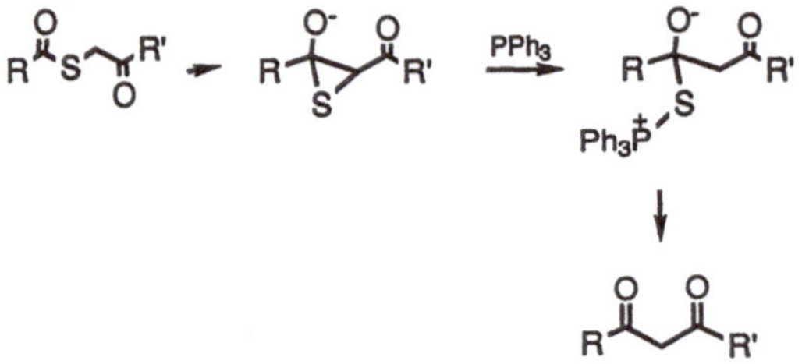

$$ \text{(191)} $$

Bridgedhead migration during Baeyer-Villiger reacton of N-carbobenzoxy-2-aza-bicyclo-[2.2.2]octan-5-one [218] is expected, as its donor character is reinforced by the hetero atom. On the other hand, an ester group at C-3 completely suppresses the influence of the nitrogen atom [219].

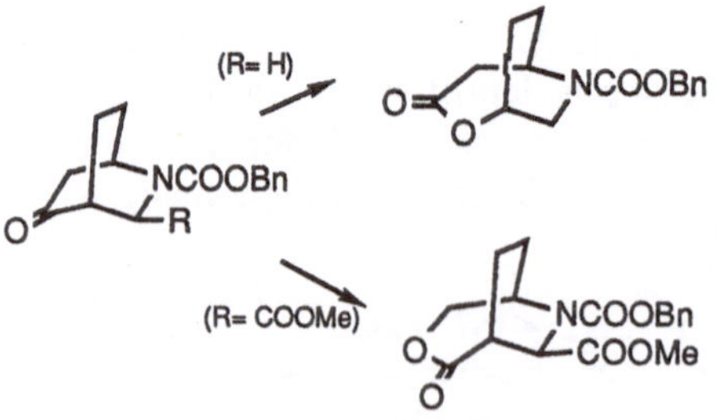

$$ \text{(192)} $$

The subtle electronic effect of a γ-substituent is also evident in another bicyclic system [220]. In this case the substituent is located outside a bridgehead and free-rotating.

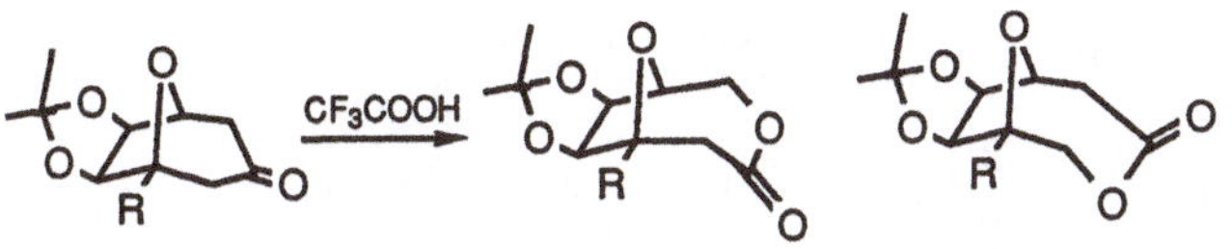

(193)

R=CH₂OSO₂CF₃	86	14
R=CH₂OBn	52	48
R=CH₃	47	53
R=Ph	88	12

Beckmann rearrangement of certain cycloalkanedione dioximes [221] has revealed regioselectivity which indicates the presence of electronic interactions between the functionalities. No matter what the reason is, the products possess arrangement of amide groups which are related to one another in the polarity alternating sense (i.e. conjointment).

(194)

(195)

(196)

The Smiles-type rearrangement of phenyl o-tolyl sulfone [222] has been shown to proceed via an addition-elimination pathway by isolation of the tricyclic intermediate. Both sites of deprotonation and addition are controlled by the sulfonyl group with contrapolarization at the second stage.

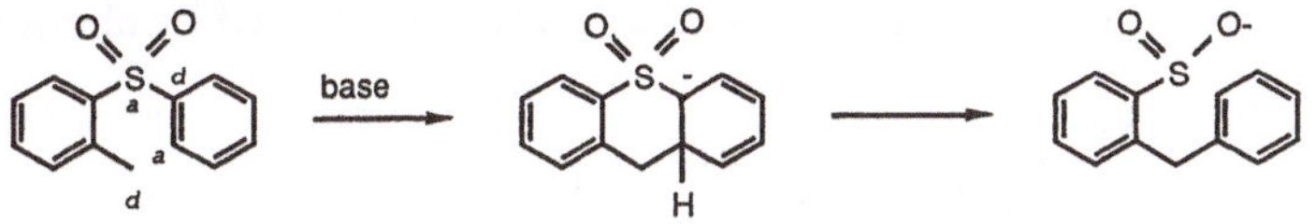

(197)

5.2 Sigmatropic Rearrangements

The Cope rearrangement is strongly accelerated when a polar group is placed at the allylic position of the 1,5-diene. Interestingly, both donors [223] and acceptors [224] are effective activators.

(198)

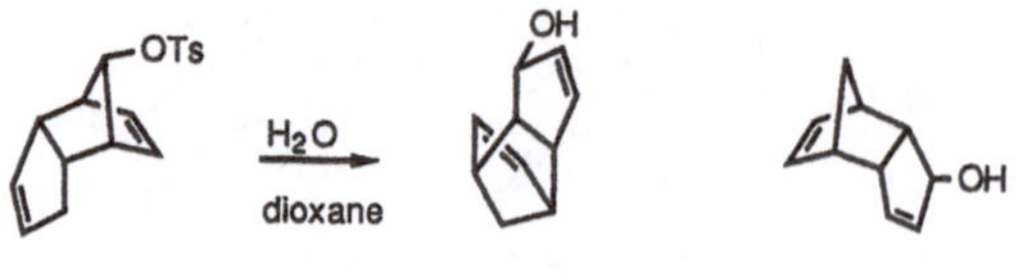

$$(199)$$

Pd(II)-catalyzed Cope rearrangement [225] occurs at room temperature, via chair-like transition states. A plausible mechanism is cyclization-induced rearrangement. Both the addition and fragmentation steps are assisted by the introduction and removal of, respectively, conjoint substituents. (Cf. Hg-salt catalyzed 1,3-transposition of allylic trichloracetimidates [226]; for other transition metal-catalyzed Clasien rearrangement, see Ref. [227].

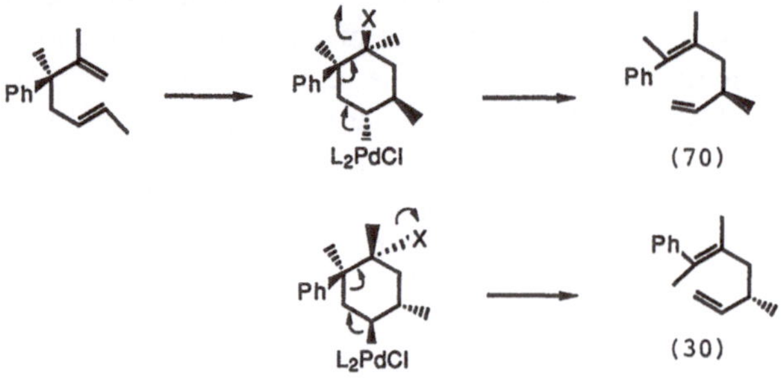

$$(200)$$

An extreme case of the Cope rearrangement proceeds via dissociation to an ion-pair. [228]. Complementary donor and acceptor stabilization of the ionic species is the cause for this abnormal reaction.

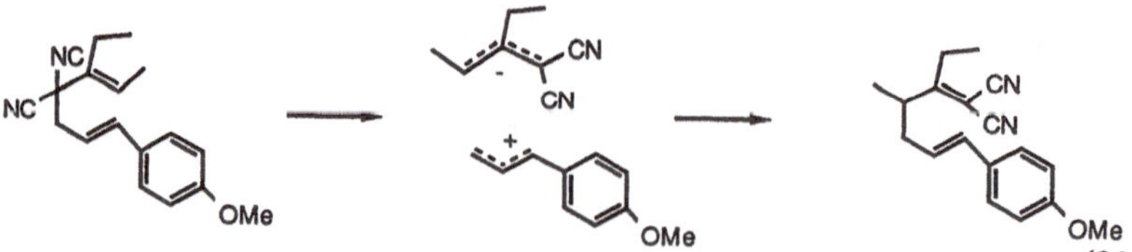

$$(201)$$

The Claisen rearrangement also benefits from a geminal donor to the framework oxygen atom [229]. This forms the basis of many synthetic applications. A newer version of the Claisen rearrangement is the carbanion-accelerated process [230]. It is noted that in all these cases the facilitation is always achieved by alternating polarity accentuation.

$$(202)-(207)$$

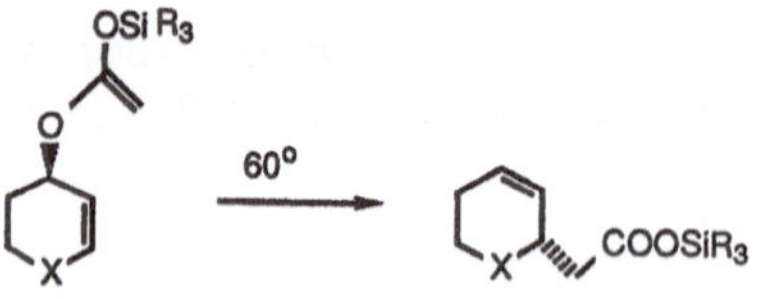

$k[o] \approx 10\text{-}25 \times k[CH_2]$

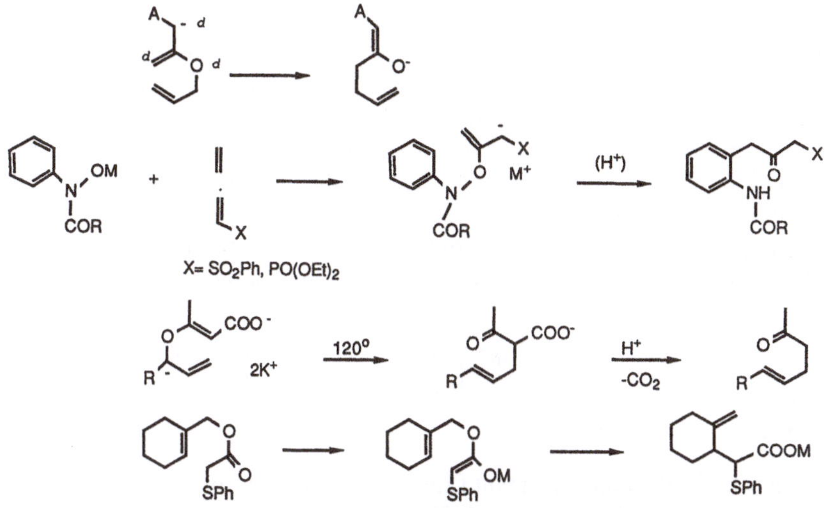

X= SO₂Ph, PO(OEt)₂

An apparent anomaly is the extremely efficient rearrangement of the enolates of α-allyloxy ketones [231]. It must be remembered that the disjoint enediolates suffer from an adverse electronic disposition and they would seek any relief.

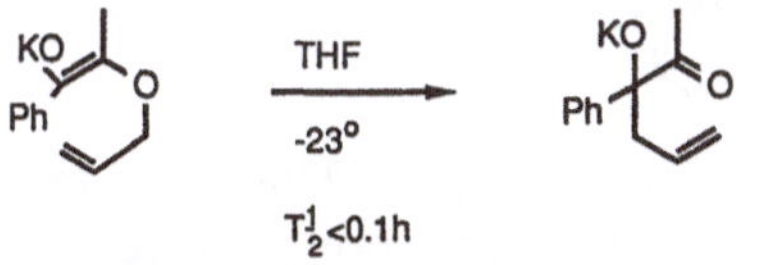

$$T_{\frac{1}{2}} < 0.1h \tag{208}$$

A biochemically significant Claisen rearrangement is the transformation of chorismate into prephenate [232] via a chair transition state. Although it is impossible to settle the question of the direction of electron flow during the reorganization, that shown by the arrows in the formula should be preferred when the influence of the various substituents is considered.

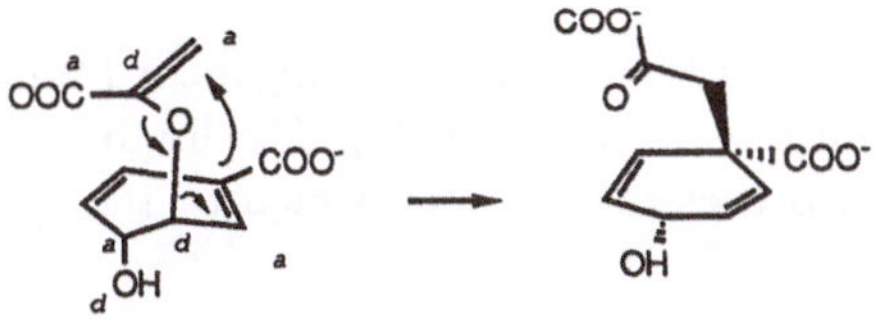

$$\tag{209}$$

Concerning the aromatic Claisen rearrangement certain salient observations are summarized in the following equations [233].

$$\tag{210}$$

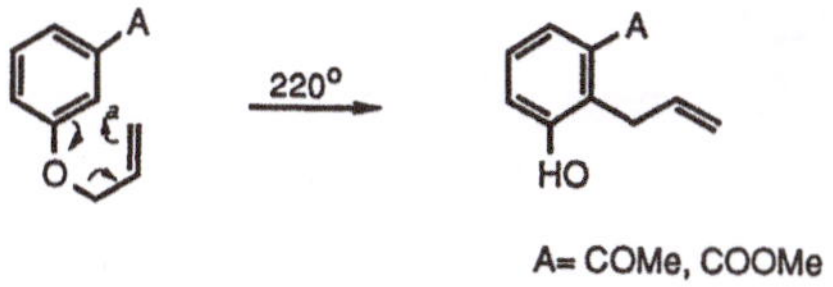

D= OMe

$$\tag{211}$$

A= COMe, COOMe

The *meta* substituent of the allyl aryl ether determines the major reaction terminus. Thus, a donor group favors the formation of 2,5-disubstituted phenol because in the transition state for the alternative sigmatropic rearrangement, allylic strain due to interference of the nuclear substituent and the CH$_2$ group develops. The rearrangement of the "electron-deficient" species might resemble nucleophilic aromatic substitution in terms of electron redistribution, and the sigma bonding occurs preferentially at the carbon closest to the acceptor group.

There are exceptions to this trend, as shown in the indole series [234]. The reason for this might be that the allyl migration to C-5 would forfeit all the aromatic stabilization.

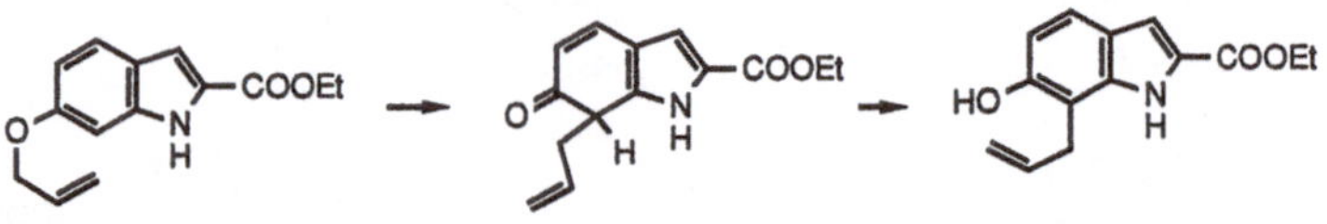

(212)

A 45-fold increase in the rearrangement rates of allyl vinyl sulfoxides relative to the corresponding sulfides has been observed [235]. The transition state for the former compounds is more polar.

(213)

Cope rearrangement of arylazo-*t*-allylmalononitriles [236] shows rate acceleration by acceptor substituents in the phenyl ring. Again, a more polar transition state favor the reorganization.

(214)

The enormous utility of a tandem aza/oxy-Cope rearrangement-Mannich cyclization is readily hinted by a neat synthesis of crinine [237]. The fragmentation as well as well as the C-C bond formation processes are orchestrated by the ideally situated hetero atoms.

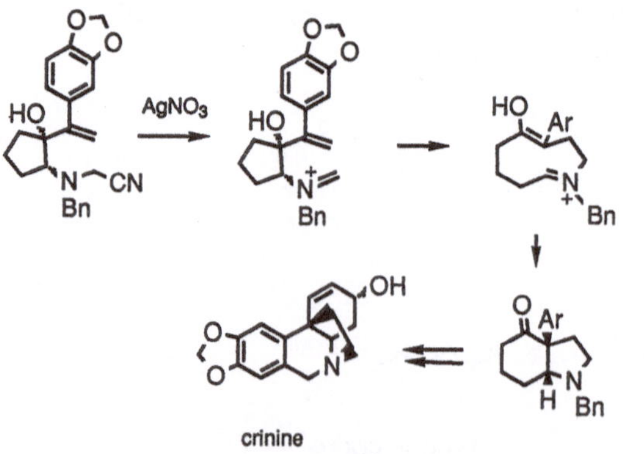

(215)

crinine

140

In an elegant approach to gelsemine [238] based on a similar strategy the cationic aza-Cope rearrangement was superseded by a simple Mannich reaction because the twistane skeleton is more highly strained. However, the desired transformation is achievable using the anionic version.

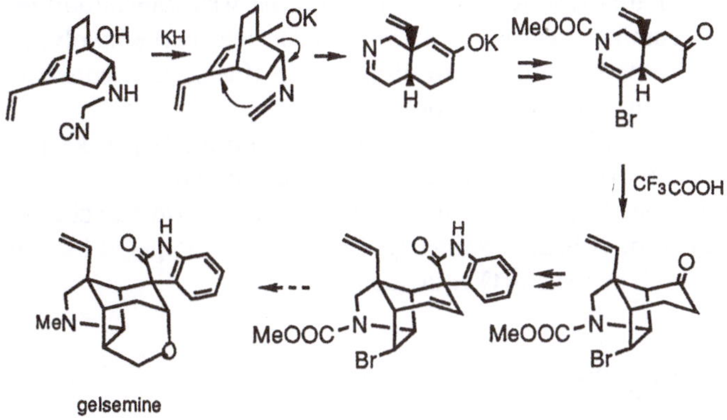

(216)

gelsemine

Another aza-Cope rearrangement which serves to elaborate a key intermediate of aspidospermine [239] may be mentioned. The distribution of functional groups facilitates the rearrangement as well as the subsequent cyclization.

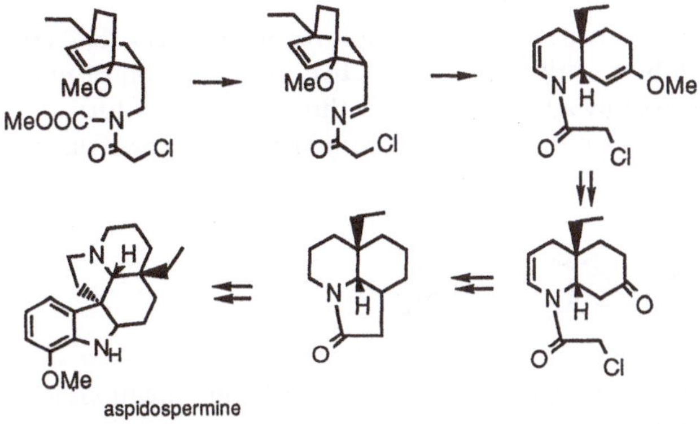

(217)

aspidospermine

The degree of regioselectivity in the rearrangement of 2-methyl-7anti-cyclopentenyl-norborn-2-en-7syn-ol [240] is perhaps unanticipated. The C-2 methyl group is responsible for the predominant generation of the allyl anion on which it has minimal electronic interaction.

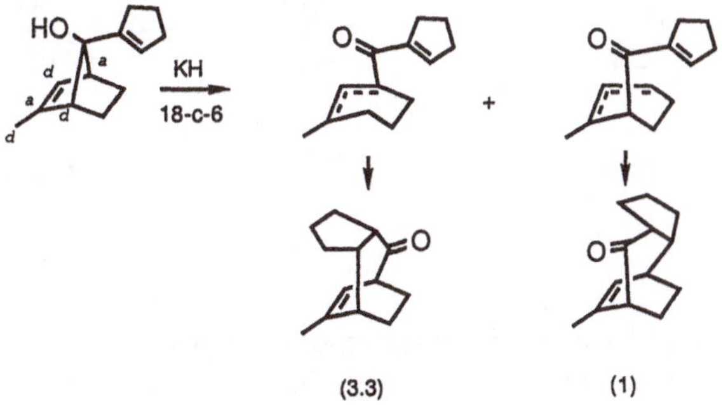

(218)

(3.3)   (1)

141

6 Redox Reactions

6.1 Oxidations

As a group the redox reactions are probably least well understood mechanistically. However, one common feature to all redox processes is the involvement of contrapolarization. Selected examples are discussed in this section concerning the effects of polar substituents on redox reactions.

The ability of percarboxylic acids to transfer the terminal hydroxyl group as "HO$^+$" to relatively weak donors such as alkenes is due to activation by the carbonyl via polarity alternation. As expected, other types of peracids with an acceptor heteroatom replacing the carbonyl behave similarly. Perhaps unanticipated by many chemists is the usefulness of Si as the C=O equivalent [241].

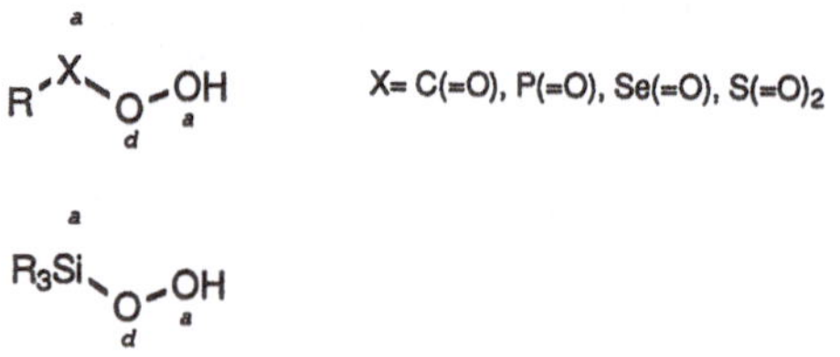

X= C(=O), P(=O), Se(=O), S(=O)$_2$

While hydrogen peroxide cannot be employed to epoxidize C,C double bonds, its combination with boron trifluoride is effective. Accentuation of polar character (in this case an acceptor) by an external agent through complex formation achieves the activation. Similar activation [242] of hydroperoxides by vanadium and titanium cations is now well known.

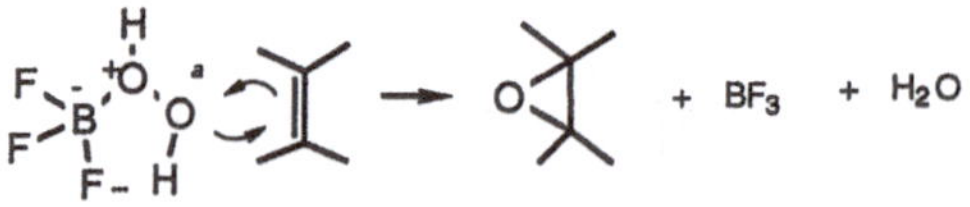

$$+ \ BF_3 \ + \ H_2O \qquad (219)$$

Epoxidation of silylallenes [243] shown in the following equations is very interesting. In each case the unique direction of the cyclopropanone opening is governed by the silyl group which is conjoint with the carbonyl.

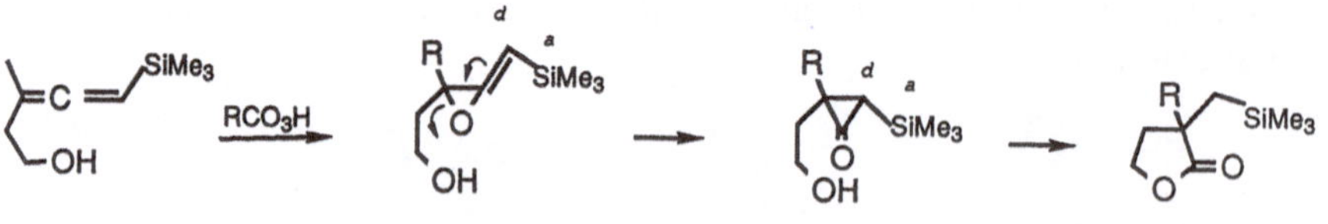

(220)

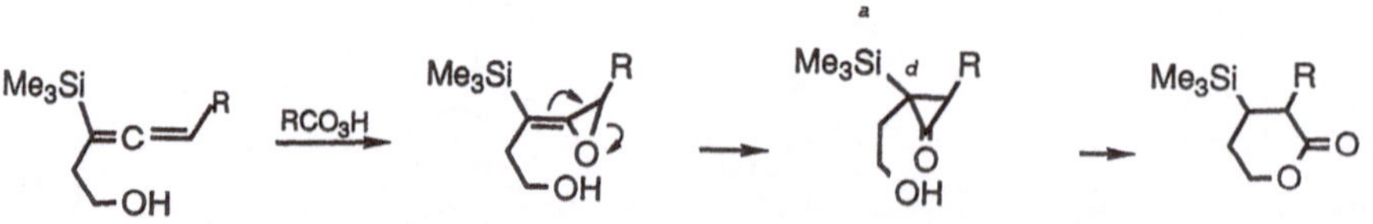

(221)

The difficult oxidation of α-oxy carbinols to α-oxy aldehydes has been experienced by many chemists [244]. One of the reasons might be the accentuation of the *a-a* interaction of the connecting carbon atoms. The situation worsens in the products.

The synthesis of α,β-unsaturated carbonyl compounds and nitriles by Pd-catalyzed reaction [245] of allyl β-oxo esters and allyl α-cyano esters is an oxidative process. With the contra-polarizability of the metal ion, elimination of a hydride from the β-position is electronically favorable.

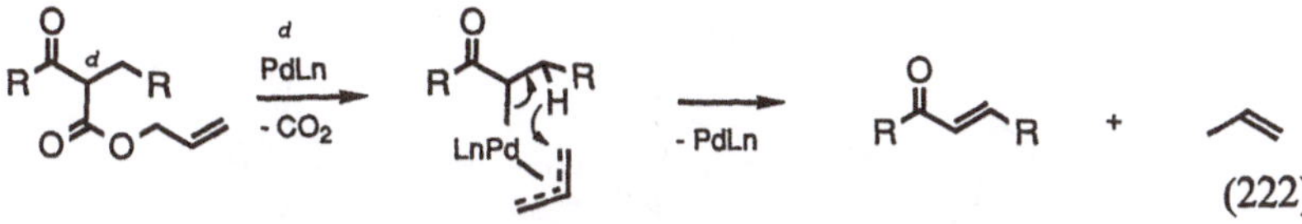

$$(222)$$

Ozonation of alkenes involves a cycloaddition-cycloreversion sequence. A synthetically significant differentiation of an alkene is shown below [246].

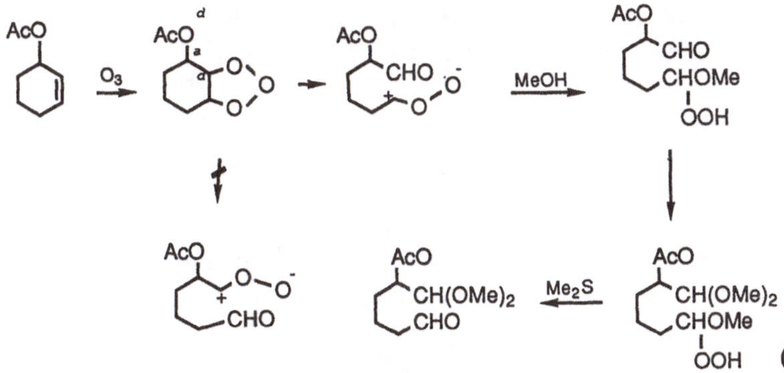

$$(223)$$

The electronic perturbation at the carbon of the primary ozonide by a proximal acetoxy group is that cycloreversion which gives rise to an intermediate with a more pronounced acceptor at the β position is disfavored. This effect is also manifested in 5-endo-acetoxynorborene, albeit in an attentuated fashion.

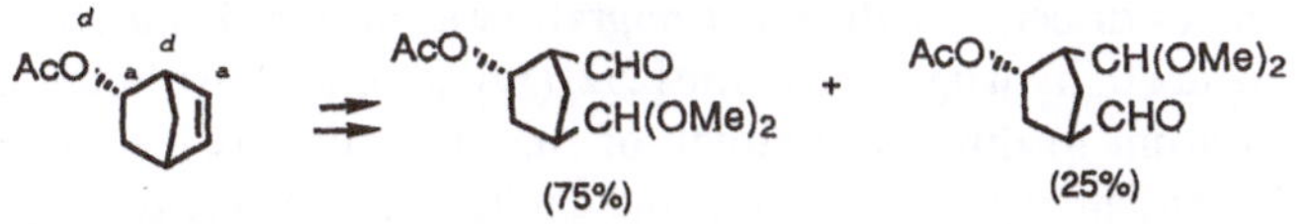

$$(224)$$

(75%) (25%)

Note the donor group is now one carbon further apart and its influence on the nearest reaction center is in the opposite sense (polarity alternation). However, a through-space electronic interaction must be important also, as the exo isomer displays no selectivity.

The original investigators noted that the preferred mode of cleavage for 5-endo-acetoxynorbornene is opposite to that predicted by the inductive effect consideration.

An application of the regiocontrol for ozonide decomposition by an allylic methoxy group (donor) is a macrolide synthesis [247].

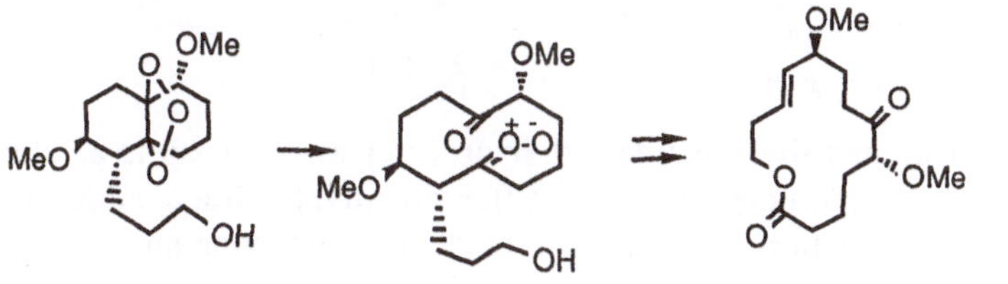

$$(225)$$

Maladehydic acid derivatives can be prepared regioselectively from 2-silylfurans by reaction with singlet oxygen [248]. Contrapolarization at the peroxy oxygen geminal to the silicon is involved, and ultimately the oxygen at the far end of the Si–C–O–O segment activates the Si–C bond cleavage. A carboxyl group serves the same purpose [249] as the silyl group.

$$\text{(226)}$$

$$\text{(227)}$$

The Baeyer-Villiger oxidation of 3-trimethylsilylcyclohexanone [250] is regioselective as indicated. The reason for the selectivity is electronic which is revealed by an examination of the relationship between the polar groups of the ketone. Here we find a disjoint sequence, the α and β carbon atoms are both quasidonors as dictated by their respective neighbors. The observed rearrangement, but not the alternative mode, relieves the unfavorable electronic interactions.

$$\text{(228)}$$

Many syntheses of natural products [251] have emerged taking advantage of this newly discovered electronic control.

It is interesting to discern the different migratory aptitudes displayed by spirocyclic ketones on exposure to hydrogen peroxide [252] (Eq. 226, 227). Although participation of the selenium atom to direct the attack of the peroxide has been formulated, the results are also consistent with an electronic explanation. Thus, the rapid formation of selenoxide renders the spirocyclic center acceptor-like through polarity alternation, and the migration of the methylene group becomes more favorable.

$$\text{(229)}$$

$$\text{(230)}$$

Oxidation of an α-silyl selenide with hydrogen peroxide leads to an aldehyde, silanol and selenol as the primary products [253]. Evidently, the fragmentation proceeds only after contrapolarization at Se, the intermediate now having an *a-d-a* array.

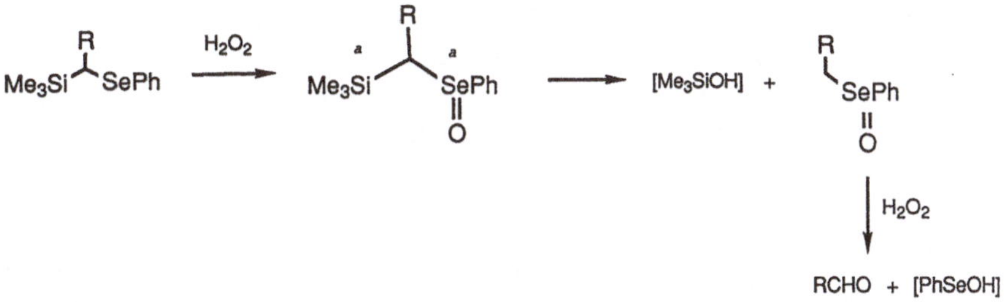

$$(231)$$

The Wesseley acetoxylation of phenols is under regiocontrol when a sulfonyl group is present in one of the *ortho* positions [254]. The same *ortho* carbon, imbued with considerable donor character, is more reluctant to receive the nucleophile AcO⁻.

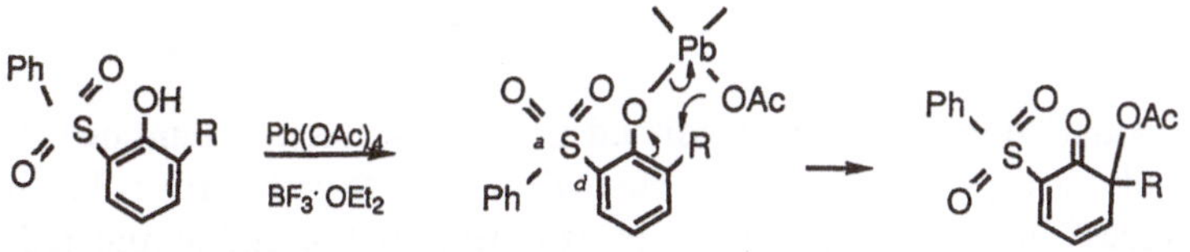

$$(232)$$

Anodic oxidation of a carboxylic acid generates a carbenium ion at the α-carbon. This contrapolarization enables a facile fragmentation of γ-keto acids [255].

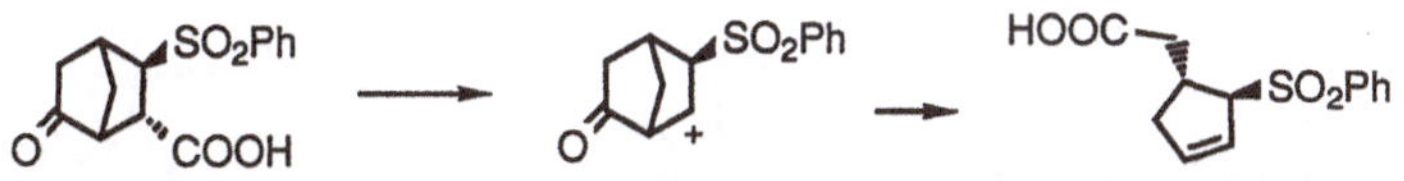

$$(233)$$

In a brilliant synthesis of ginkgolide B [91], it was planned to construct the tetra-hydrofuran ring by a remote oxidation of a tertiary alcohol with a lead tetraacetate-iodine combination. However, an isomeric ring structure emerged. The non-occurrence of the desired reaction may be due to the resistent ethereal oxygen which discourages its α-carbon to assume a donor role. Although the reaction is essentially homopolar, it seems that a slight degree of polar characteristics has an enormous effect. In this case a negative influence on a proximal atom diverts the reaction site.

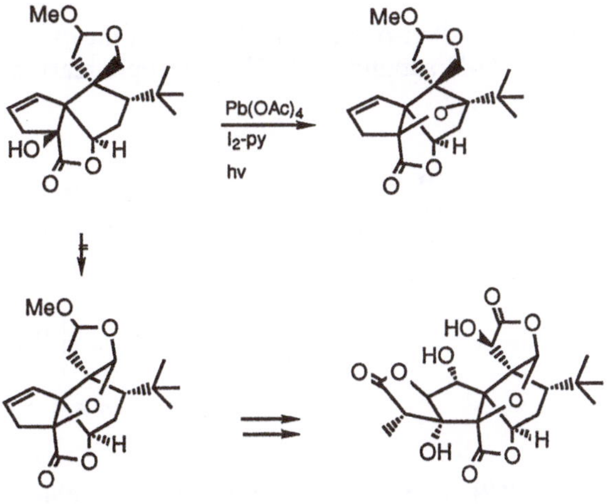

$$(234)$$

ginkgolide B

Despite the mechanistic obscurity of C-H bond oxidation by chromic acid, regioselectivity has been discerned in well defined systems [256]. For example, oxidation of *endo*-fenchyl acetate and the bornyl acetates (*exo* and *endo*) gives ketones in which the new carbonyl group is derived from a donor carbon.

(235)

(236)

The oxidative deamination mediated by 3,5-di-*t*-butyl-*o*-quinone [257] could very well involve a C → O 1,5-sigmatropic hydrogen shift within the Schiff base network. This process in essence accomplishes oxidation of the amine and reduction of the quinone. The interesting point is the strong donor oxygen forces the nitrogen atom into an acceptor role during the reaction.

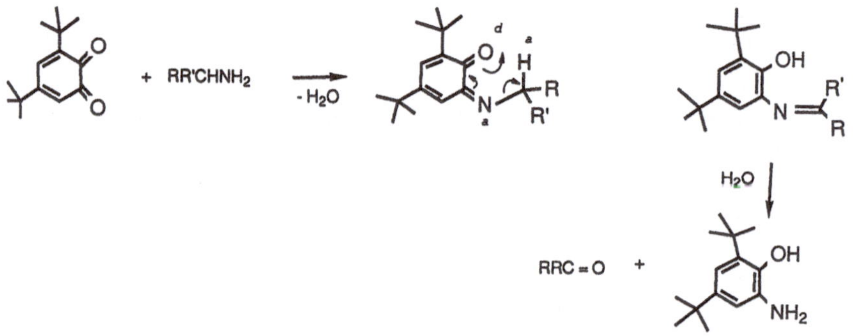

(237)

Oxidative demethylation of *N*-alkyl-*N*-methylanilines with nitrone [258] proceeds via intermediates having a N-O-N subunit. A [2.3]sigmatropic rearrangement leads to the decomposition.

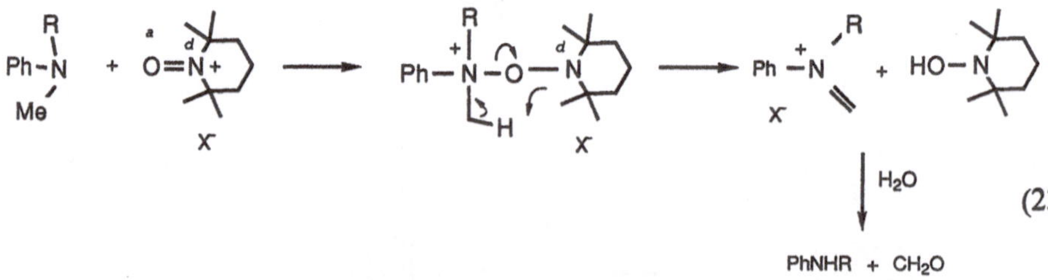

(238)

A synthesis of glaziovine [259] via oxidative phenol-aniline coupling involves contrapolarizing activation of the aromatic carbon para to the amino group.

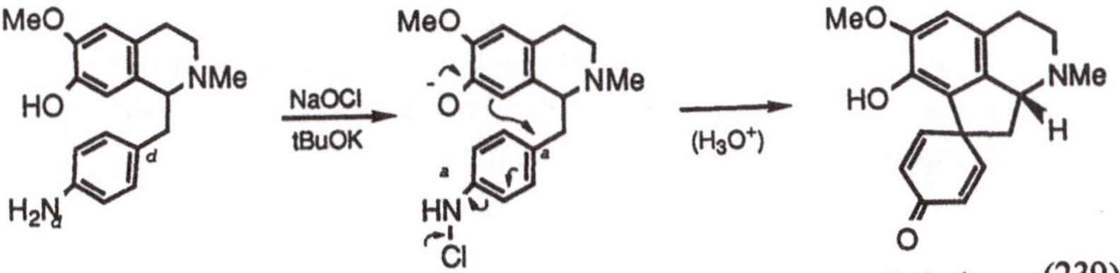

glaziovine   (239)

Other types of remote activation are illustrated in the syntheses of kreysigine [260] and *N*-methyllaurotetanine [261].

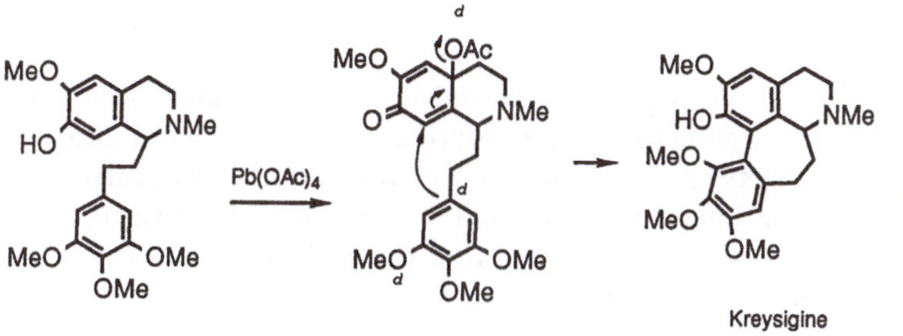

(240)

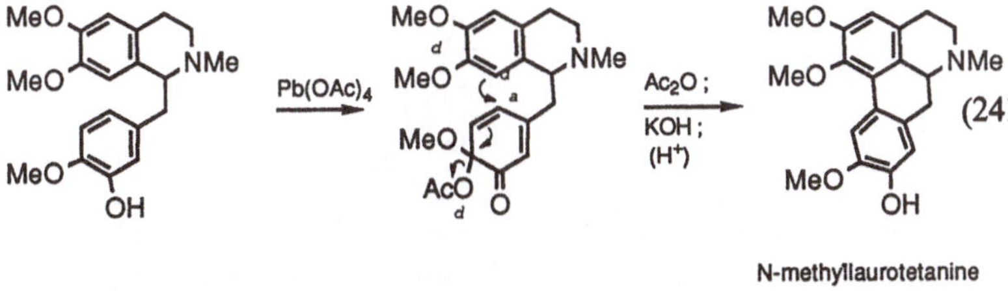

Kreysigine

(241)

N-methyllaurotetanine

An unexpected rearrangement of a 1,3-bishomocubyl ring system [262] accompanies an attempted nitration. The oxidative conditions are requisite for the transformation of a 1,3-dinitro substitution pattern to one with a disjoint polarity sequence (contra-polarization involved). The unusual contrathermodynamic reaction is allowed to occur because of the special steric confinement of the functionalized carbon atoms. The initial fragmentation and Michael addition are also set up by the proper disposition of the nitro groups.

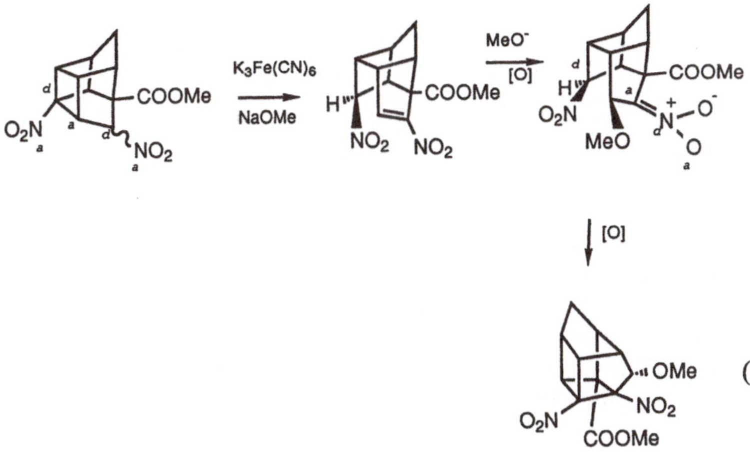

(242)

147

Fremy's salt has been used to degrade a *p*-hydroxybenzylamine [263]. Aminoxylation at the *para* position of the phenol triggers fragmentation; the C-O bond formation gives rise to a conjoint 1,3-disubstituted system, despite contrapolarization at the intervening oxygen atom is required.

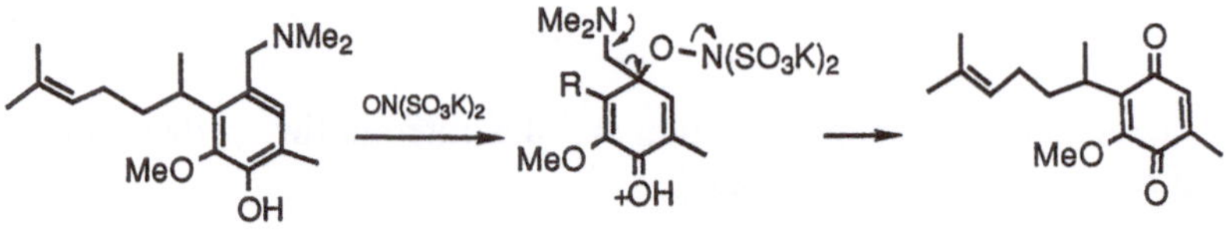

$$(243)$$

Succinic acids undergo bisdecarboxylation on exposure to lead tetraacetate [264]. Contra-polarization at one of the α-carbons through fragmentation of the lead(IV) carboxylate moiety enables a smooth decarboxylation of the remaining functionality.

In the presence of a juxtapositional double bond such succinic acids could form dilactones [265]. The ethereal oxygen of the lead(IV) carboxylate intermediate acts as an acceptor in this instance.

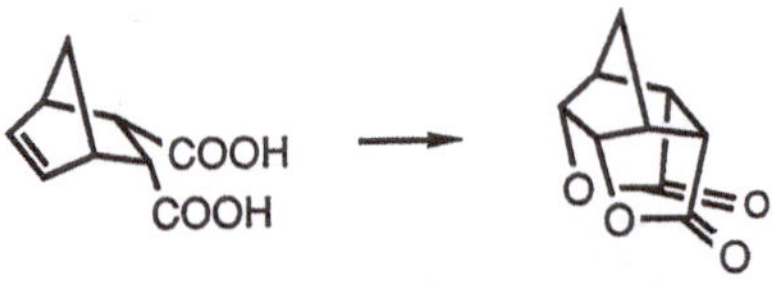

$$(244)$$

Protection of amino compounds by the 2,4-bis(methylthio)phenoxycarbonyl residue [266] converts them into acid stable entities. Removal of the protecting group can be achieved through oxidation at sulfur such that the resulting bissulfone provides conjoint pathways for ready fragmentation.

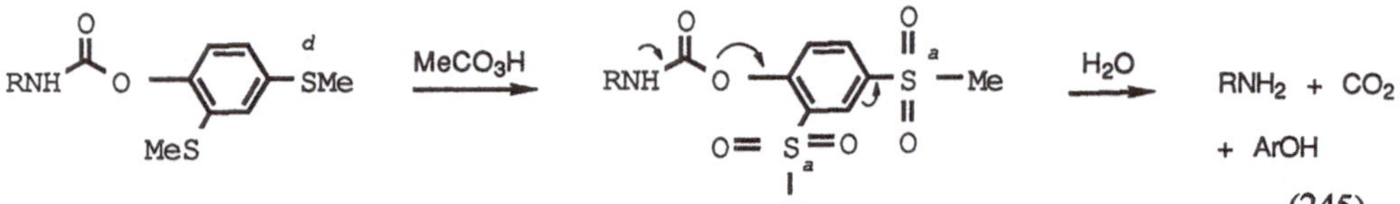

$$(245)$$

6.2 Reduction

The facilitation of the C-halogen bond reduction by a neighboring carbonyl group is due to the imposed donor character of the α-carbon, and by extension, the heightened acceptor character of the halogen atom.

There is a net reduction of α-chlorosulfenyl-α-chloroketones under hydrolytic conditions [267]. The sulfur atom undergoes a series of contrapolarizing changes (*a-d-a*). The carbonyl group also plays a crucial accentuation role in the various steps of the reaction.

148

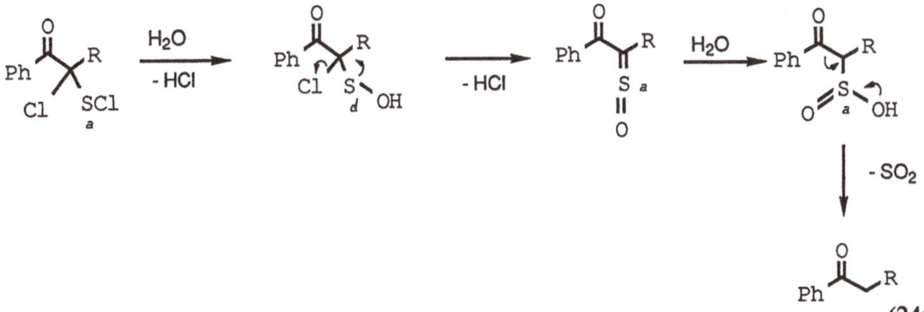

(246)

Hydride reduction of 3,3,5-trimethylcyclohexanone and its 5-substituted analogues exhibits a rate dependence on the substituent [268]. There is rate enhancement and rate retardation, respectively, with an acceptor and a donor. Enhancement is due to the existence of a destabilizing d-d bonding situation between the α- and β-carbon atoms, retardation might reflect a reluctance of the molecule to diminish a polarity alternating sequence.

(247)

	$10^3 k_{ax}$	$10^3 k_{eq}$
X=CN	15	235
X=H	1.5	11.5
X=F	0.7	5.3

Regioselective reduction of a succinimide intermediate for a (+)-heliotridine synthesis [269] must be due to the influence of the acetoxy group. It is conjectured that the a-a array of the intervening carbon atoms activates the carbonyl. Steric effects should be minimal on the choice of the two sites.

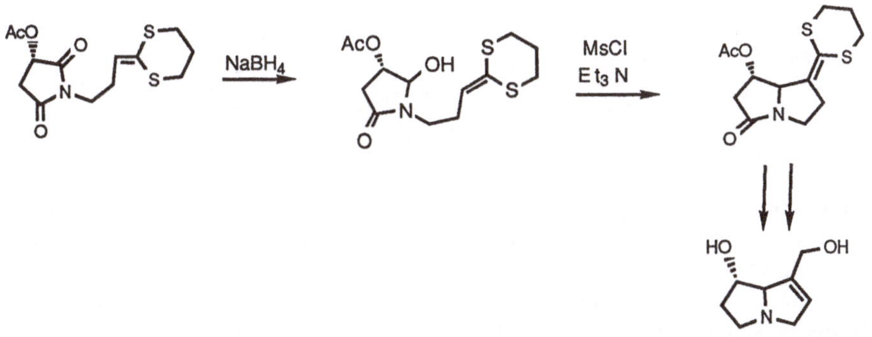

(248)

It is tempted to speculate the contribution of the remote polar substituent to the regioselective reduction of the endione intermediates employed for the syntheses of tetrodotoxin [270] and reserpine [271]. Without disregarding the steric hindrance of the substituent in each case to the proximal carbonyl, the disjoint relationship of which to the substituent is noted. The increased donor character of this proximal carbonyl may also be translated into a lesser reactivity.

(249)

(250)

2,2-Difluorocyclopropyl esters behave quite differently from other halo analogs under solvolytic conditions [272]. The formation of 2,2-difluorocarbonyl compounds which emerges as an important pathway, may be considered as an internal redox process. The diminished significance of the concerted dehalogenative ring opening is due to the high strength of the C-F bond and the stabilization of the carbanion by the fluorine atoms via a polarity alternation mechanism.

(251)

It may be difficult to imagine the alkylation of a 3,5-dialkoxyphenylcarbanion in which no ortho chelators are available to assist deprotonation at the desired position, not to mention the disjoint relationship of this donor carbon to the oxy substituents present. A solution to this problem is provided by redox manipulations and incorporation of a removable acceptor group to affirm the donor site. Thus, the Birch reduction product of 3,5-dimethoxybenzoic acid readily undergoes alkylation, and oxidative decarboxylation generates the aromatic compound [273].

(252)

Two points are noteworthy. Birch reduction (and all other redox processes) always affords umpoled species; restoration of the original state must involve a reverse reaction (i.e. oxidation).

Perhaps it should be mentioned also the orientation of the Birch reduction which is strongly dependent on the nature of the aromatic substituents. Donor-substituted benzenes furnish predominantly 1-substituted 1,4-cyclohexadienes while acceptor-substituted analogues give 3-substituted 1,4-cyclohexadienes. The regioselectivities can be explained by the destabilizing d-d pairing in the intermediates from d-substituted cyclohexadienyl radical anions leading to the 3-substituted products, and the stabilizing effect of an *a-d* pairing in the *a*-substituted intermediates of the same pattern.

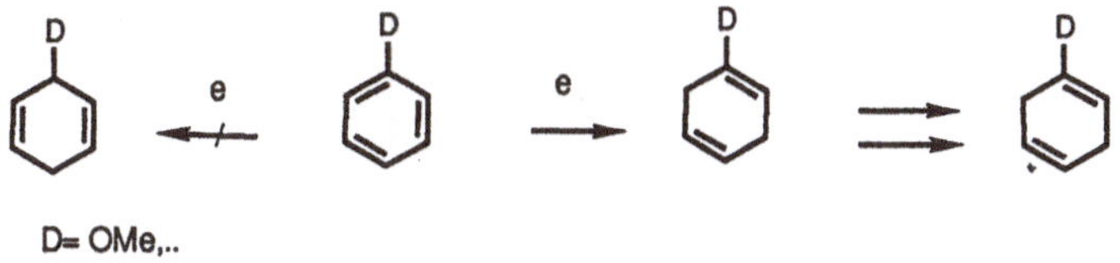

$$D = OMe,..$$ (253)

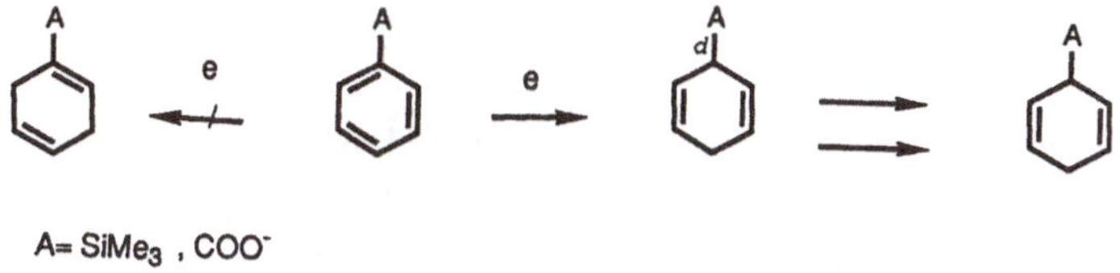

$$A = SiMe_3, COO^-$$ (254)

As a further illustration of the regiochemistry associated with the Birch reduction of anisole derivatives, 11β-acetoxy-*O*-methylestrone [274] offers an interesting case.

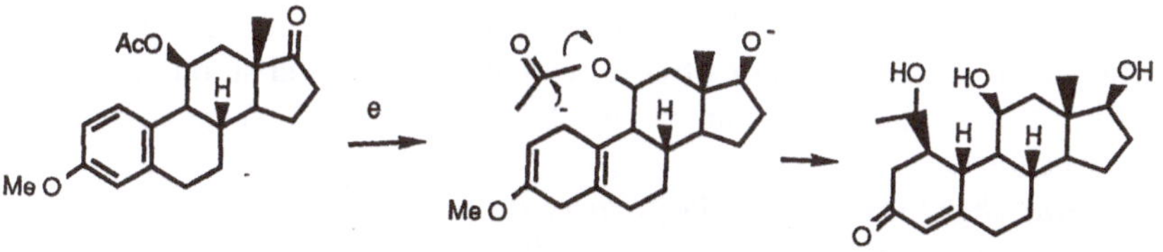

(255)

Cleavage of alkyl aryl ethers with sodium or potassium metal shows excellent regioselectivity [275]. Using limiting radical anion structures to analyze the results and noting the preference for such structures as dependent on the substituents, it is possible to formulate reaction pathways leading to the observed products.

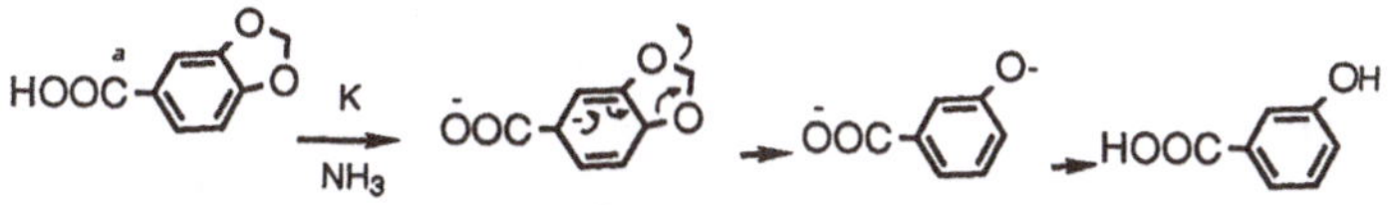

(256)

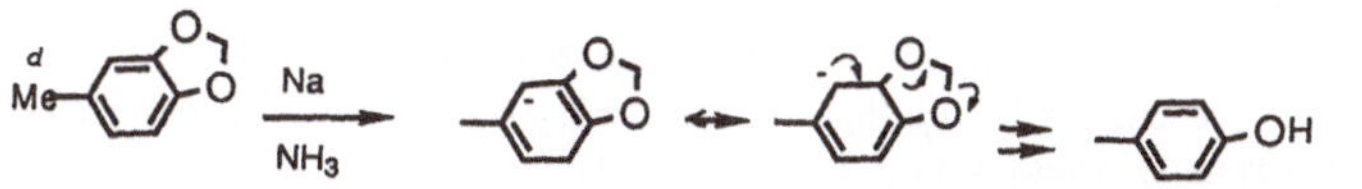

(257)

It should be emphasized that, in piperonic acid, for example, the leaving oxygen is conjoint with the dominating acceptor (COO), and fragmentation as shown can only occur when contrapolarization intervenes.

Perhaps another set of reactions showing the pronounced effect of contrapolarization pertaining to redox processes should be discussed. In the following equations the relative rates for ether cleavage are indicated. Note the undesirable positioning of the ethereal oxygen in the two substrates. In *p*-nitrophenyl benzyl ether particularly, heterolysis is against natural polarity alternation, and only through contrapolarizing conditions is the alkyl C-O bond scission possible, albeit with great sluggishness.

151

$$O_2N-\underset{a}{\bigcirc}-CH_2\overset{d}{OPh}\quad\xrightarrow[k_{rel=1}]{}\quad O_2N-\bigcirc-CH_2\overset{-}{}\quad+\ PhO^{-}\qquad(258)$$

$$O_2N-\underset{a}{\bigcirc}-\overset{d}{O}\diagdown_{CH_2}Ph\quad\xrightarrow[k_{rel=10^{-4}}]{}\quad O_2N-\bigcirc-\overset{-}{O}\quad+\ PhCH_2\qquad(259)$$

7 Conclusions

In view of the extensive documentation outlined above, the usefulness of the polarity alternation concept as a primary guide for evaluation of substituent effects can hardly be denied. The influence of a substituent on the *ipso* site has not been discussed in this article but an even more direct and important effect is implicit. Among the innumerable examples one may cite the preferential formation of geminal dimetallic species [5] in hydrometalation and carbometalation of vinylmetals and acetylenes. On the other hand, chemical systems are usually very complex, inter- and intramolecular forces including steric and stereoelectronic factors may dominate over polarity alternation. Thus, chelation by a proximal donor often directs metalation and stabilizes certain organometallic entities. In these instances the stability gaining from polarity alternation is overwhelmed.

The sensible approach to employing the polarity alternation concept is to be flexible. When applied correctly to a reactivity problem the failure of PAR should elicit a deeper and more detailed analysis. New facets of an organic system may be revealed.

In a previous article [8] a brief review of bifunctional chains of atoms with respect to their reactivities and modes of formation was presented. Consequently, the polarity concept is of some value to systematization of organic reactions and it should be helpful to planning synthesis.

8 References

1. Ho T-L (1988) Rev. Chem. Intermed. 9: 117
2. Pople JA, Gordon M (1967) J. Am. Chem. Soc. 89: 4253
3. Wiberg KB, Breneman CM, Laidig KE, Rosenberg RE (1989) Pure Appl. Chem. 61: 635
4. Gray AP, Kraus H (1966) J. Org. Chem. 31: 399
5. Klein J (1988) Tetrahedron 44: 503
6. Adams SR, Kao JPY, Grynliewicz G, Minta A, Tsien RY (1988) J. Am. Chem. Soc. 110: 3212
7. Apeloig Y, Karni M: J. Chem. Soc. Perkin II 1988: 625
8. Ho T-L (1989) Res. Chem. Intermed. 11: 157
9. Jacquesy J-C, Jouantannetaud M-P, Makani S: Chem. Commun. 1980: 110
10. Maslak P, Fanwick PE, Guthrie RD (1984) J. Org. Chem. 49: 655
11. Burnell RH, Jean M, Poirier D (1988) Can. J. Chem. 65: 775; Burnell RH, Cote C (1988) Synth. Commun. 18: 1753
12. For a discussion of *ipso* ion formation during aromatic nitration, see Myhre PC (1976) ACS Symp. Ser. 22: 87
13. Miller RB, Tsang T (1988) 3rd N. Am. Chem. Congr. ORGN 184
14. Iwata C, Yamada M, Shiono Y (1979) Chem. Pharm. Bull. 27: 274

15. Tsuge O, Kanemasa S, Naritomi T, Tanaka J: Chem. Lett. 1984: 1255
16. Semmelhack MF, Clark G, Farina R, Saeman M (1979) J. Am. Chem. Soc. 101: 217
17. Semmelhack MF, Yamashita A (1980) J. Am. Chem. Soc. 102: 5924
18. Boutinnet J-C, Rose-Munch F, Rose E (1985) Tetrahedron Lett. 26: 3989
19. Sutherland RG, Chowdhury RL, Piorko A, Lee CC (1986) Can. J. Chem. 64: 2031
20. Gilday JP, Widdowson DA: Chem. Commun. 1986: 1235
21. Keumi T, Hamanaka K, Hasegawa H, Minamide N, Inoue Y, Kitajima H: Chem. Lett. 1988: 1285
22. Markosza M, Winiarski J (1987) Acc. Chem. Res. 20: 282
23. Bellas M, Suschitzky H: Chem. Commun. 1965: 367
24. Biehl ER, Patrizi R, Reeves PC (1971) J. Org. Chem. 36: 3252
25. Pansegrau PD, Rieker WF, Meyers AI (1988) J. Am. Chem. Soc. 110: 7178
26. Lourak M, Vanderesse R, Fort Y, Caubere P (1988) Tetrahedron Lett. 29: 545
27. Vanelle P, Jentzer O, Bahnous M, Crozet MP (1988) Tetrahedron Lett. 29: 5361
28. Bordwell FG, Wilson CA (1987) J. Am. Chem. Soc. 109: 5470
29. Dainter RS, Suschitzky H, Wakefield BJ, Hughes N, Nelson AJ, J.C.S. Perkin I 1988: 227
30. Raju B, Krishna Rao GS Synthesis 1987: 197
31. Viehe HG, Janousek Z, Merenyi R, Stella L (1985) Acc. Chem. Res. 18: 148
32. Wenkert E (1984) Heterocycles 21: 325
33. Williams RM, Dung JS, Josey J, Armstrong R, Meyers H (1983) J. Am. Chem. Soc. 105: 3214
34. Page PCB, Rosenthal S, Williams RV: Synthesis 1988: 621
35. Furuya S, Okamoto T (1988) Heterocycles 27: 2609
36. Parker RE, Isaacs NS (1959) Chem. Rev. 59: 737
37. Sharpless KB, Behrens CH, Katsuki T, Lee AWM, Martin VS, Takatani M, Viti SM, Walker FJ, Woodard SS (1983) Pure Appl. Chem. 55: 589
38. McManus MJ, Berchtold GA, Jerina DM (1985) J. Am. Chem. Soc. 107: 2977
39. Ayi AI, Remli M, Guedy R (1981) Tetrahedron Lett. 22: 1505
40. Corey EJ, Melvin LS, Jr, Haslanger MF: Tetrahedron Lett. 1975: 3117
41. Fuchs PL (1976) J. Org. Chem. 41: 2937
42. Alexakis A, Jachiet D (1988) Tetrahedron Lett. 29: 217
43. White DR: Tetrahedron Lett. 1976: 1753
44. Baldwin JE (1976) Chem. Commun. 734: 738
45. Satoh T, Iwamoto K, Yamakawa K (1987) Tetrahedron Lett. 28: 2603
46. Evans DA, Biller SA (1985) Tetrahedron Lett. 26: 1907
47. Gao Y, Sharpless KB (1988) J. Am. Chem. Soc. 110: 7538
48. McGarvey GJ, Williams JM, Hiner RN, Matsubara Y, Oh T (1986) J. Am. Chem. Soc. 108: 4943
49. Gladysz JA (1988) 196th ACS Nat. Meet. ORGN 210
50. Glass RS: Chem. Commun. 1971: 1546; Hendrickson JB, Bergeron R, Giga A, Sternbach DD (1973) J. Am. Chem. Soc. 95: 3412
51. Ho T-L, Wong CM: Chem. Commun. 1973: 224; (1973) Synth. Commun. 3: 145
52. Hendrickson JB, Bergeron R: Tetrahedron Lett. 1973: 4607
53. Höfle G, Steglich W, Vorbrüggen H (1978) Angew. Chem. Int. Ed. Engl. 17: 569
54. Barton DHR, Lacher B, Zard SZ (1988) Tetrahedron 43: 4321
55. Belleau B, Malek G (1968) J. Am. Chem. Soc. 90: 1651
56. Mukaiyama T (1976) Angew. Chem. 88: 111
57. Sheehan JC, Hess GP (1955) J. Am. Chem. Soc. 77: 1067
58. Bodanszky M (1984) Principles of peptide synthesis, Springer, Berlin Heidelberg, New York, p 168
59. Kametani T, Umezawa O (1964) Chem. Pharm. Bull. 12: 379
60. Aoki T, Haga N, Sendo Y, Konoike T, Yoshioka M, Nagata W (1985) Tetrahedron Lett. 26: 339
61. Ho T-L, Wong CM (1974) J. Org. Chem. 39: 3453
62. Büchi G, Kulsa P, Ogasawara K, Rosati RL (1970) J. Am. Chem. Soc. 92: 999
63. Kozar LG, Clark RD, Heathcock CH (1977) J. Org. Chem. 42: 1386
64. Tsuji J, Yuhara M, Minato M, Yamada H, Sato F, Kobayashi Y (1988) Tetrahedron Lett. 29: 343

65. Heck RF (1979) Acc. Chem. Res. 12: 146
66. Trost BM, Verhoeven TR (1982) In Wilkinson G, Stone FGA (ed) Comprehensive organo-metallic chemistry, vol 8 Pergamon, Oxford
67. Chan TH, Mychajlowskij W: Tetrahedron Lett. 1974: 3479
68. Lee JG, Bartsch RA (1979) J. Am. Chem. Soc. 101: 228
69. Baciocchi E, Ruzziconi R, Sebastiani GV (1983) J. Am. Chem. Soc. 105: 6114
70. Ho T-L (1989) J. Chem. Educ. 66: 785
71. Qian C-P, Nakai T (1988) Tetrahedron Lett. 29: 4119
72. Swenton JS, Jurcak JG (1988) J. Org. Chem. 53: 1530
73. Lambert JB, Wang G-T, Finzel RB, Teramura DH (1987) J. Am. Chem. Soc. 109: 7838
74. Hendrickson JB, Sternbach DD (1975) J. Org. Chem. 40: 3450
75. Katritzky AR, Fan W-Q, Li Q-L (1987) Tetrahedron Lett. 28: 1195
76. Moody CJ, Warellow GJ (1987) Tetrahedron Lett. 28: 6089
77. Evans DA, Britton TC (1987) J. Am. Chem. Soc. 109: 6881
78. Hicks HB, Harris DW, Feather MS, Loeppky RN (1974) J. Agri. Food Chem. 22: 724
79. Hicks HB, Feather MS (1977) Carbohydr. Res. 54: 209
80. Corey EJ, Mitra RB, Uda H (1964) J. Am. Chem. Soc. 86: 485
81. Dastur KP (1974) J. Am. Chem. Soc. 96: 2505
82. Greenlee ML (1981) J. Am. Chem. Soc. 103: 2425
83. Corey EJ, De B (1984) J. Am. Chem. Soc. 106: 2735
84. Winkler JD, Muller CL, Scott RD (1988) J. Am. Chem. Soc. 110: 4831
85. Bartlett PD, Ando T (1970) J. Am. Chem. Soc. 92: 7518
86. Stach H, Hesse M (1986) Helv. Chim. Acta 69: 1614
87. Deno NC, Sacher E (1965) J. Am. Chem. Soc. 87: 5120
88. Grob CA, von Tschammer H (1968) Helv. Chim. Acta 51: 1083
89. Sternbach D, Shibuya M, Jaisli F, Bonetti M, Eschenmoser A (1979) Angew. Chem. 91: 670
90. Bredereck H, Effenberger F, Simchen G (1963) Chem. Ber. 96: 1350
91. Corey EJ (1988) Chem. Soc. Rev. 17: 111
92. Kuehne ME, Roland DM, Hafter R (1978) J. Org. Chem. 43: 3705; Kuehne ME, Matsko TH, Bohnert JC, Kirkemo CL (1979) J. Org. Chem. 44: 1063
93. Stork G, Dolfini JE (1963) J. Am. Chem. Soc. 85: 2872
94. Hugel G, Gourdier B, Levy J, LeMen J: Tetrahedron Lett. 1974: 1597
95. Kuehne ME, Podhorez DE (1985) J. Org. Chem. 50: 924
96. Natsume M, Utsunomiya I (1984) Chem. Pharm. Bull. 32: 2477
97. Moore HW (1979) Acc. Chem. Res. 12: 125, and refs. therein
98. Marshall JA, Buse CT, Seitz DE (1973) Synth. Commun. 3: 85; Trost BM, Rigby J (1976) J. Org. Chem. 41: 3217
99. Inokuchi T, Takagishi S, Akahoshi F, Torii S: Chem. Lett. 1987: 1553
100. Wenkert E (1980) Acc. Chem. Res. 13: 27
101. Marino JP, Laborde E (1985) J. Am. Chem. Soc. 107: 734
102. Marino JP, Long JK (1988) J. Am. Chem. Soc. 110: 7916
103. Guyot M, Molho D: Tetrahedron Lett. 1973: 3433
104. Posner GH, Asirvatham E, Webb KS, Jew SS (1987) Tetrahedron Lett. 28: 5071
105. Sommer LH, Bailey DL, Goldberg GM, Buck CE, Bye TS, Evans FJ, Whitmore FC (1954) J. Am. Chem. Soc. 76: 1613
106. Fleming I, Lawrence NJ (1988) Tetrahedron Lett. 29: 2073
107. Akers JA, Bryson TA (1989) Tetrahedron Lett. 30: 2187
108. Evans DA, Fu GC, Hoveyda AH (1988) J. Am. Chem. Soc. 110: 6917
109. Liotta D, Zima G, Saindane M (1982) J. Org. Chem. 47: 1258
110. Chamberlin AR, Dezube M, Dussault P, Mills MC (1983) J. Am. Chem. Soc. 105: 5819
111. Seebach D (1988) 196th ACS Nat. Meet. ORGN 257
112. Krow GR, Fan DM (1974) J. Org. Chem. 39: 2674
113. Krow GR, Shaw DA, Lynch, B, Lester W, Szczepanski SW, Raghavachari R, Derome AE (1988) J. Org. Chem. 53: 2258
114. Carrupt PA, Vogel P (1982) Tetrahedron Lett. 23: 2563; Black KA, Vogel P (1986) J. Org. Chem. 51: 5341
115. Gates M, Tschudi G (1956) J. Am. Chem. Soc. 78: 1380

116. Raunio EK, Frey TG (1971) J. Org. Chem. 36: 345
117. Truce WE, Tichenor GJW (1972) J. Org. Chem. 37: 2391; Sasaki T, Kojima A, Ohta M: J. Chem. Soc. [C] 1971: 196
118. Petrov ML, Petrov AA (1972) Zh. Oshch. Khim. 42: 2345; (1973) 43: 691
119. Brandsma L, Bos HJT, Arens JF (1968) In Viehe HG (ed) The chemistry of acetylene, Dekker, New York, Chap. 11
120. Kwok WK, Lee WG, Miller SI (1969) J. Am. Chem. Soc. 91: 468
121. Schroth W, Peschel J, Zschunke A (1969) Z. Chem. 9: 110
122. Freear J, Tipping AE: J. Chem. Soc. [C] 1969: 411
123. Overman LE, Sharp MJ (1988) J. Am. Chem. Soc. 110: 612
124. Büchi G, Wüest H (1979) J. Org. Chem. 44: 546
125. Magnus P, Schultz J, Houk KN (1986) Tetrahedron Lett. 27: 655
126. Armstrong RJ, Harris FE, Weiler L (1982) Can. J. Chem. 60: 673
127. Schmid R, Huesmann PL, Johnson WS (1980) J. Am. Chem. Soc. 102: 5122; Johnson WS, Ward CE, Boots SG, Gravestock MB, Markezich RL, McCarry BE, Okorie DA, Parry RJ (1981) J. Am. Chem. Soc. 103: 88
128. Yoshioka H, Takasaki K, Kobayashi M, Matsumoto T: Tetrahedron Lett. 1979: 3489
129. Stevens RV, Kenney PM (1983) J. Am. Chem. Soc. 105: 384
130. Corey EJ, Boaz NW (1985) Tetrahedron Lett. 26: 6019
131. Lipshutz BH, Ellsworth EL, Siahaan TJ (1988) J. Am. Chem. Soc. 110: 4834
132. Danheiser RL, Fink DM (1985) Tetrahedron Lett. 26: 2509
133. Ficini J, Revial G, Genet JP (1981) Tetrahedron Lett. 22: 629, 633
134. Yoshida K, Yamamoto M, Ishigura M: Chem. Lett. 1986: 1059
135. Krohn K, Priyuno W (1986) Angew. Chem. Intern. Ed. 25: 339
136. Singh LW, Ila H, Junjappa H: Synthesis 1987: 873
137. Kono M, Saitoh Y, Shirahata K, Arai Y, Ishii S (1987) J. Am. Chem. Soc. 109: 7224
138. Pandit UK, Bieraugel H, Stoit AR (1984) Tetrahedron Lett. 25: 1513; Ando M, Büchi G, Ohnuma T (1975) J. Am. Chem. Soc. 97: 6880
139. Baldwin JE, Li CS: Chem. Commun. 1988: 261
140. Stevens RV, Pruitt J: Chem. Commun. 1983: 1425
141. Marino JP, Mesbergen WB (1974) J. Am. Chem. Soc. 96: 4050
142. Barros MT, Geraldes CFGC, Maycock CD, Silva MI (1988) Tetrahedron 44: 2283
143. Wieland T, Birr C, Wissenbach H (1969) Angew. Chem. 81: 782
144. Adam W, Rücktaschel R (1972) J. Org. Chem. 37: 4128
145. Adam W, Liu JC, Rodriguez O (1973): J. Org. Chem. 38: 2269
146. Utimoto K, Kato S, Tanaka M, Hoshino Y, Fujikura S, Nozaki H (1982) Heterocycles 18: 149
147. Badejo IT, Mamanta MT, Fry JL (1988) 196th ACS Nat. Meet. ORGN 12
148. Rao YK, Nagarajan M (1988) Tetrahedron Lett. 29: 107
149. Sommer LH, Pioch RP (1954) J. Am. Chem. Soc. 76: 1606
150. Smith AB III (1989) Pure App. Chem. 61: 405
151. Corey EJ, Balanson RD (1974) J. Am. Chem. Soc. 96: 6516
152. Benson W, Winterfeldt E (1979) Chem. Ber. 112: 1913
153. Overman LE, Mendelson LT, Jacobson EJ (1983) J. Am. Chem. Soc. 105: 6629
154. Hiemstra H, Speckamp WN (1988) In: Brossi A (ed.) The alkaloids, vol 32 chap 4, Academic, New York
155. Cushman M, Dekow FW (1978) Tetrahedron 34: 1435
156. Sanchez IH, Soria JJ, Lopez FJ, Larraza MI, Flores HJ (1984) J. Org. Chem. 49: 157
157. Takano S, Murakata C, Ogasawara K (1981) Heterocycles 16: 247
158. Masamune S, Ang SK, Egli C, Nakatsuka N, Sarkar SK, Yasunari Y (1967) J. Am. Chem. Soc. 89: 2506
159. Lukanov LK, Venkov AP, Mollov NM: Synthesis 1987: 204
160. Hartwig U, Pritzkow H, Sundermeyer W (1988) Chem. Ber. 121: 1435
161. Allmann R, Hanefeld W, Krestal M, Spangenberg B (1987) Angew. Chem. Inter. Ed. 26: 1133
162. Khanapure SP, Biehl ER (1988) 196th ACS Nat. Meet. ORGN 69
163. Kessar SV, Singh M: Tetrahedron Lett. 1969: 1155
164. Wakamatsu T, Miyachi N, Ozaki F, Shibasaki M, Ban Y (1988) Tetrahedron Lett. 29: 3829
165. Krabbenhoft HO (1978) J. Org. Chem. 43: 1305

166. Fliri A, Hohenlohe-Oehringer K (1980) Chem. Ber. 113: 607
167. Ficini J, Guigant A, d'Angelo J (1979) J. Am. Chem. Soc. 101: 1318
168. Sasaki T, Kanematsu K, Uchide M: Tetrahedron Lett. 1971: 4855
169. Dennis N, Katritzky A, Parton SK: J. Chem. Soc. Perkin I 1976: 2285
170. Oppolzer W, Grayson JI, Wegmann H, Urrea M (1983) Tetrahedron 39: 3695
171. Burger K, Neuhauser H, Eggersdorfer M: Synthesis 1987: 924
172. Furuta K, Miwa Y, Iwanaga K, Yamamoto H (1988) J. Am. Chem. Soc. 110: 6254
173. Gassman PG, Singleton DA, Wilwerding JJ, Chavan SP (1987) J. Am. Chem. Soc. 109: 2182
174. Wulff WD, Yang DC (1983) J. Am. Chem. Soc. 105: 6726
175. Vedejs E, Eberlein TH, Mazur DJ, McClure CK, Perry DA, Ruggeri R, Schwartz E, Stults JS, Varie DL, Wilde RG, Wittenberger S (1986) J. Org. Chem. 51: 1556
176. Jung ME, Lowe JA: Chem. Commun. 1978: 95
177. Tou JS, Reusch W (1980) J. Org. Chem. 45: 5012
178. Watt DS, Corey EJ: Tetrahedron Lett. 1972: 4651
179. Kametani T, Suzuki T, Ichikawa Y, Fukomoto K: J. Chem. Soc. Perkin I 1975: 2102
180. Marazano C, LeGoff MT, Fourrey JL, Das BC: Chem. Commun. 1981: 389
181. Chou S-SP, Sun D-J: Chem. Commun. 1988: 1176
182. Overman LE, Petty CB, Ban T, Huang GT (1983) J. Am. Chem. Soc. 105: 6335
183. Vedejs E, Miller WH, Pribish JR (1983) J. Org. Chem. 48: 3611
184. Avenati M, Carrupt P-A, Quarroz D, Vogel P (1982) Helv. Chim. Acta 65: 188
185. Demarchi B, Vogel P, Pinkerton AA (1988) Helv. Chim. Acta 71: 1249
186. Vogel P (1983) Kagaku Zokan 99: 21
187. Büchi G, Mak CP (1977) J. Am. Chem. Soc. 99: 8073
188. Hoffman R, Wells P, Morrison H (1971) J. Org. Chem. 36: 102
189. Chapman OL, Smith HG, King PW (1963) J. Am. Chem. Soc. 85: 806
190. Corey EJ, Bass JD, LeMahieu R, Mitra RB (1964) J. Am. Chem. Soc. 86: 5570
191. Wender PA, Lechleiter JC (1978) J. Am. Chem. Soc. 100: 4321
192. Yanagiya M, Kaneko K, Kagi T, Matsumoto T: Tetrahedron Lett. 1979: 1761
193. Fraser-Reid B, Hicks DR, Walker DL, Iley DE, Yunker MB, Tam SY-K, Anderson RC: Tetrahedron Lett. 1975: 297
194. Mattay J, Gersdorf J, Freudenberg U (1984) Tetrahedron Lett. 25: 817
195. Krafft ME (1988) J. Am. Chem. Soc. 110: 968
196. Nazarov IN, Torgov IB, Terekhova LN: Izv. Akad. Nauk SSSR, Otd. Khim. Nauk 1942: 200
197. Denmark SE, Jones TK (1982) J. Am. Chem. Soc. 104: 2642
198. LePerchec P: Synthesis 1975: 1
199. Oppolzer W, Thirring K (1982) J. Am. Chem. Soc. 104: 4978
200. Carpenter BK (1978) Tetrahedron 34: 1877
201. Fitjer L, Quabeck U (1987) Angew. Chem. Intern. Ed. 26: 1023
202. Morrow GW, Wang S, Swenton JS (1988) Tetrahedron Lett. 29: 3441
203. Glover SA, Goosen A, McCleland CW, Schoonraad JL (1987) Tetrahedron 43: 2577
204. Giese B, Groeninger KS, Witzel T, Korth H-G, Sustmann R (1987) Angew. Chem. Intern. Ed. 26: 233
205. Woodward RB (1968) Pure Appl. Chem. 17: 519
206. Stork G, Kowalski C, Garcia G (1975) J. Am. Chem. Soc. 97: 3258
207. Magnus P, Moerck R (unpublished results) quoted in Wilkinson G, Stone FGA (ed) (1982) Comprehensive organometallic chemistry, vol 7 Pergamon, Oxford, p 519
208. Brownbridge P, Fleming I, Pearce A, Warren S: Chem. Commun. 1976: 751
209. Tanino K, Katoh T, Kuwajima I (1988) Tetrahedron Lett. 29: 1815
210. Durst T, Tin K-C: Tetrahedron Lett. 1970: 2369
211. Shamma M, Jones CD (1969) J. Am. Chem. Soc. 91: 4009; (1970) 92: 4943; Shamma M. Nugent JF (0973) Tetrahedron 29: 1265
212. Harley-Mason J, Kaplan M: Chem. Commun. 1967: 915
213. Kende AS (1960) Org. React. 11: 261
214. Paquette LA (1977) Org. React. 25: 1
215. Fuji K, Usami Y, Sumi K, Ueda M, Kajiwara K: Chem. Lett. 1986: 1655
216. Wallach O (1918) Liebigs Ann. Chem. 414: 233
217. Roth M, Dubs P, Gotschi E, Eschenmoser A (1971) Helv. Chim. Acta 54: 710

218. Krow GR (1981) Tetrahedron 37: 2697
219. Baxter AJG, Holmes AB: J. Chem. Soc. Perkin I 1977: 2343
220. Noyori R, Sato T, Kobayashi H: Tetrahedron Lett. 1980: 2569
221. Iwakura Y, Uno K, Haga K, Nakamura K (1973) J. Polym. Sci., Polym. Chem. Ed. 11: 367; Hall HK Jr (1958) J. Am. Chem. Soc. 80: 6404; Glover GI, Smith RB, Rapoport H (1965) J. Am. Chem. Soc. 87: 2003
222. Drozd VN, Frid YY (1967) Zh. Org. Khim. 3: 373
223. Evans DA, Golob AM (1975) J. Am. Chem. Soc. 97: 4765
224. Breslow R, Hoffman JM (1972) J. Am. Chem. Soc. 94: 2111
225. Overman LE, Jacobsen EJ (1982) J. Am. Chem. Soc. 104: 7225
226. Overman LE (1974) J. Am. Chem. Soc. 96: 597
227. Schenck TG, Bosnich B (1985) J. Am. Chem. Soc. 107: 2058
228. Wigfield DC, Feiner S, Taymaz K: Tetrahedron Lett. 1972: 891, 895
229. Ireland RE, Mueller RH (1972) J. Am. Chem. Soc. 94: 5897; Curran DP, Suh YG (1984) J. Am. Chem. Soc. 106: 5002
230. Denmark SE, Harmata MA (1982) J. Am. Chem. Soc. 104: 4972; Blechert S (1984) Tetrahedron Lett. 25: 1547; Baldwin JE, Tzodikov NR (1977) J. Org. Chem. 42: 1878; Büchi G, Vogel DE (1985) J. Org. Chem. 50: 4664
231. Koreeda M, Luengo JI (1985) J. Am. Chem. Soc. 107: 5572
232. Copley SD, Knowles JR (1985) J. Am. Chem. Soc. 107: 5306
233. Bruce JM, Roshan-Ali Y: J. Chem. Soc. Perkin I 1981: 2677
234. Moody CJ: J. Chem. Soc. Perkin I 1984: 1333
235. Block E, Ahmad S (1985) J. Am. Chem. Soc. 107: 6731; Hwu JR, Anderson DA (1986) Tetrahedron Lett. 27: 4965
236. Mitsuhashi T (1986) J. Am. Chem. Soc. 108: 2400
237. Overman LE, Jacobsen EJ (1982) Tetrahedron Lett. 23: 2741
238. Earley WG, Jacobsen EJ, Meier GP, Oh T, Overman LE (1988) Tetrahedron Lett. 29: 3781; Earley WG, Oh T, Overman LE (1988) Tetrahedron Lett. 29: 3785
239. Wu PL, Chu M, Fowler FW (1988) J. Org. Chem. 53: 963
240. Paquette LA, Pierre F, Cottrell CE (1987) J. Am. Chem. Soc. 109: 5731
241. Ho T-L (1979) Synth. Commun. 9: 37
242. Sharpless KB, Verhoeven TR (1979) Aldrichim. Acta 12: 63
243. Bertrand M, Dulcere JP, Gil G (1980) Tetrahedron Lett. 21: 1945
244. Parkes KEB, Pattenden G: J. Chem. Soc. Perkin I 1988: 1119
245. Minami I, Nisar M, Yuhara M, Shimizu I, Tsuji J: Synthesis 1987: 992
246. Schreiber SL, Claus RE, Reagan J (1982) Tetrahedron Lett. 23: 3867
247. Schreiber SL, Liew W-F (1985) J. Am. Chem. Soc. 107: 2980
248. Adam W, Rodriguez A (1981) Tetrahedron Lett. 22: 3505
249. White JD, Carter JP, Kezar HS (1982) J. Org. Chem. 47: 929
250. Hudrlik PF, Hudrlik AM, Nagendrapper G, Yimenu T, Zellers ET, Chin E (1980) J. Am. Chem. Soc. 102: 6894
251. Asaoka M, Shima K, Takei H (1987) Tetrahedron Lett. 28: 5669; Asaoka M, Takenouchi K, Takei H, (1988) Tetrahedron Lett. 29: 325; Asaoka M, Fujii N, Shima K, Takei H: Chem. Lett. 1988: 805
252. Trost BM, Buhlmayer P, Mao M (1982) Tetrahedron Lett. 23: 1443
253. Sachdev K, Sachdev HS: Tetrahedron Lett. 1976: 4223
254. Quinkert G, Heim N, Glenneberg J, Billhardt U-M, Autze V, Bats JW, Dürner G (1987) Angew. Chem. Intern. Ed. 26: 362
255. Michaelis R, Mueller U, Schaefer HJ (1987) Angew. Chem. Intern. Ed. 26: 1026
256. Freeman F (1986) In: Mijs WJ, deJonge CRHI (ed) Organic syntheses by oxidation with metal compounds, Plenum, New York, chap. 2
257. Corey EJ, Achiwa A (1969) J. Am. Chem. Soc. 91: 1429
258. Hunter DH, Racok JS, Rey AW, Ponce YZ (1988) J. Org. Chem. 53: 1278
259. Kametani T, Takahashi K, Ogasawara K, Fukumoto K: Tetrahedron Lett. 1973: 4219
260. Hoshino O, Toshioka T, Umezawa B: Chem. Comm. 1972: 740
261. Hare H, Hashimoto F, Hoshino O, Umezawa B (1984) Tetrahedron Lett. 25: 3615

262. Marchand AP, Jin P-W, Flippen-Anderson JL, Gilardi R, George C: Chem. Commun. 1987: 1108
263. Saa JM, Llobera A (1987) Tetrahedron Lett. 28: 5045
264. Sheldon RA, Kochi JK (1972) Org. React. 19: 279
265. Alder K, Schneider S (1936) Liebigs Ann. Chem. 524: 189
266. Kunz H, Lasowski H-J (1986) Angew. Chem. 98: 170
267. Oka K, Hara S: Tetrahedron Lett. 1977: 695
268. Agami C, Kazakos A, Levisalles J, Sevin A (1980) Tetrahedron 36: 2977
269. Chamberlin AR, Chung JYL (1983) J. Am. Chem. Soc. 105: 3653
270. Kishi Y, Aratani M, Fukuyama T, Nakatsubo F, Goto T, Inoue S, Tanino H, Sugiura S, Kakoi H (1972) J. Am. Chem. Soc. 94: 9217, 9219
271. Woodward RB, Bader FE, Bickel H, Frey AJ, Kierstead RW (1958) Tetrahedron 2: 1
272. Crabbé P, Cervantes A, Cruz A, Gleazzi E, Iriarte J, Velarde E (1973) J. Am. Chem. Soc. 95: 6653; Kobayashi Y, Taguchi T, Mamada M, Shimizu H, Murahashi H (1979) Chem. Pharm. Bull. 27: 3123
273. Crombie LW, Crombie WML, Firth DF: J. Chem. Soc. Perkin I 1988: 1263
274. Magerlein BJ, Hogg JA (1958) J. Am. Chem. Soc. 80: 2220
275. Maercker A (1987) Angew. Chem. Intern. Ed. 26: 972

Author Index Volumes 151—155

Author Index Vols. 26–50 see Vol. 50
Author Index Vols. 50–100 see Vol. 100
Author Index Vols. 101–150 see Vol. 150

The volume numbers are printed in italics